# PROGRAM.
# CONTROLL...

Pete DeReamer

USTA# 2006134035
tnui2tee

# PROGRAMMABLE CONTROLLERS
## Concepts and Applications

**First Edition**

C. T. Jones
L. A. Bryan

**An IPC/ASTEC Publication**
**Atlanta**

*Library of Congress Cataloging in Publication Data*

Jones, C.T. (Clarence T.), 1953–
    Programmable controllers.

    Bibliography: p.
    Includes index.
    1. Programmable controllers.     I. Bryan, L.A.
(Luis A.), 1956–     . II. Title.
TJ223.P76J66 1983   629.8'95   83-82790
ISBN 0-915425-00-9

*This book was set in Helvetica by Regal Type*
*It was printed and bound by Phoenix Communications*

*Technical Revision by Kim Ahmed and Eric Bryan*
*Edited by Karen Newchurch*
*Illustrations and art direction by Garon Hart*

*To Janis, and to Bryan,*
*for their long-patience*

*—Clarence*

*To my parents,*
*for their continuing*
*support and encouragement*

*—Luis*

# Preface

*Knowledge is power.*
—Francis Bacon

In the thirteen years since their inception, programmable controllers have proven to be the salvation of many United States manufacturing plants which previously relied on electromechanical control systems. Unfortunately, simply because of the lack of knowledge about programmable control systems and other automation technology, many other industrial manufacturers are continuing to use outdated, less efficient production techniques. With the flood of new technology inundating our society, the amount of new information and the range of new products have become so diverse that narrowing the field and selecting the best product for the particular application have become costly and time-consuming tasks. In addition, heightened foreign competition has forced American manufacturers to seek actively the most prolific, precise, and profitable means for generating quality products.

In many cases, manufacturing problems can be solved by installing a programmable control system, such as a programmable controller. However, arranging the most effective system requires the expertise to define specific needs, select and match the appropriate components, and build the system with consideration of proper design and construction methods. Until now, the only sources of information concerning programmable controllers were users's manuals, published by manufacturers about their equipment; and technical articles in trade magazines, written by engineers at a highly technical level. These technical articles more often are concerned with theory than with application. This book — the first of its kind dedicated exclusively to programmable controllers — supplement these other sources by integrating information necessary for planning, selecting, building, and interfacing PCs in most applications.

In writing this book, we chose to assemble a wide variety of topics. By design, we include information often taken for granted, for our less technically skilled readers, as well as advanced methodology and applications, for our readers who are more familiar with PCs. In addition, the range of information covered makes the book especially well-suited for use by colleges, universities, and technical schools. Since PCs are similar functionally and operationally, we feel that readers at all levels will find this book to be a valuable reference tool and useful design and construction guide for programmable control systems.

Because each chapter of the book builds upon information presented in every preceding chapter, it is important that new users of programmable controllers comprehend early material before advancing to later chapters. A glossary, index, and appendix are provided to facilitate the usefulness of this book and to aid in understanding unfamiliar terminology or concepts.

We hope that the knowledge gained from this book will enable our readers to perform their jobs more effectively, to understand the manufacturers' literature better, to become more aware of the potential benefits of the current technology, and to make more informed choices in control systems and their applications.

# About the Authors

**Clarence Jones** graduated from Howard University with a Bachelor of Science in Electrical Engineering. During his college years, he worked in the cooperative education program with Procter and Gamble, receiving early exposure and valuable insight into the burgeoning field of programmable controllers.

In 1978, Mr. Jones began working with Proctor and Gamble as an electrical control systems engineer and was involved in all phases of their manufacturing control systems, including definition design, and start-up. In 1981, he joined a leading manufacturer of programmable controllers as an applications engineer, conducting on-site lectures and seminars, configuring specific control systems design, and trouble-shooting various system malfunctions. With the expertise gained through his previous education and experience, Mr. Jones organized International Programmable Controls, Inc. in January of 1983 as a consulting organization for industrial automation. In recognition of the need for a comprehensive guide to PCs, early in 1983, Mr. Jones began writing *Programmable Controllers* as his initial project with IPC.

Mr. Jones participates in several professional organizations including the Institute of Electrical and Electronics Engineers, the Engineering Society of Detroit, and the Instrument Society of America.


**Luis Bryan** received the Bachelor of Science in Electrical Engineering with high honors and the Master of Engineering in Electrical Engineering with highest honors, both from the University of Tennessee. While involved in his graduate studies, Mr. Bryan consulted on several projects with national governmental agencies including "Advanced Instrumentation for Reflood Studies," sponsored by the United States Department of Energy at the Oak Ridge National Laboratories. During his college career, he was named to Phi Kappa Phi honor society with Eta Kappa Nu engineering honorary.

In 1979, Mr. Bryan began working as an applications engineer with one of the major manufacturers of programmable controlles. For the next four years (in Canada, Mexico, and several South American countries), he supervised international marketing, product applications, and gave lectures and seminars on the application of PCs. Mr. Bryan has trained users, sales personnel and distributors and has defined system configurations for specific applications. In early 1983, he joined with Mr. Jones and helped to organize International Programmable Controls, Inc. and is interested in the international expansion of industrial automation.

Mr. Bryan is active in several professional organizations including the Institute of Electrical and Electronics Engineers, as well as the IEEE's Instrument Society and Computer Society, and is a senior member of the Instrument Society of America.

# Acknowledgements

This first book on the subject of programmable controllers has been made a reality by the efforts and support of many people. We are greatly indebted to the ASTEC organization for assisting us in publishing this work. An especial appreciation is extended to Garon Hart, who not only provided exceptional art direction, but also guidance through the many steps of the book making process, during our first publishing venture. A primary objective in creating this book was to provide a valuable source of information on PCs in a language that could be easily understood. After many technical revisions and constructive criticism, by Eric Bryan and Kim Ahmed, we feel that this objective was achieved. We would like to thank Karen Newchurch for the long hours spent during the final editing.

We also extend our appreciation to Zachery Senior, who contributed to the typing of the manuscript, and to James Miller for his consulting on Chapter 11. Writing a book like this would have been impossible without the cooperation made by the many manufacturers of programmable controllers and support products. We would like to thank the following companies who provided us with product photos and permission to use certain illustrations from user manuals.

Allen-Bradley
Barber-Coleman
Cincinnati Electrosystems
Cincinnati Milacron
Control Technology
Eagle Signal
Electronic Processors
General Electric
Gidding & Lewis
Gould Modicon
GTE Sylvania
Industrial Data Terminals

Klockner-Moeller
Minarik Electric
Omron Electronics
Reliance Electric
Square D
Struthers-Dunn
Teletype
Texas Instruments
Westinghouse Electric
Xcel Controls
Xicor
Xycom

# Contents

# Part III: UNDERSTANDING THE SOFTWARE COMPONENTS

# Part IV: SPECIAL TOPICS

# PART
# --]I[--

# INTRODUCTORY CONCEPTS

Through this discussion of the evolution, principles of operation, and areas of application of programmable controllers, readers should be well on their way to a basic understanding of PCs.

# CHAPTER
# --]1[--

# INTRODUCTION TO PROGRAMMABLE CONTROLLERS

*In a relatively short period of time the philosophy of industrial control has been revolutionized. Much responsibility is owed to a single product, the programmable controller.*

This chapter is devoted to discussion of the evolution of programmable controllers, their theory of operation, their application, and of the benefits of their use.

## 1-1 HISTORICAL BACKGROUND

The design criteria for the first programmable controller were specified in 1968 by the Hydramatic division of the General Motors Corporation. The primary goal was to eliminate the high cost associated with inflexible, relay-controlled systems. The specifications required a solid-state system with computer flexibility, suited to survive the industrial environment, to be easily programmed and maintained by plant engineers and technicians, and last but not least, to be reusable. Such a control system would reduce machine downtime and provide expandability for the future.

### The Early Programmable Controller

The first programmable controllers were more than just relay replacers. Because they were capable of on/off control only, their application was limited to machines and processes that required repetitive operation, such as transfer lines and grinding and boring machines. On the other hand, programmable controllers were an improvement over relays. They were easily installed, used considerably less space and energy, had diagnostic indicators that aided trouble-shooting, and unlike relays, were reusable if a project were scrapped. The initial design not only met Hydramatic requirements, but led to further improvements which would spread the use of programmable controllers to other industries.

### Early Innovations

From 1970 to 1974, early innovations in microprocessor technology added greater flexibility and intelligence to the programmable controller. Capabilities of operator interfaces, arithmetic, data manipulation, and computer communications added new dimensions to PC applications.

The Cathode Ray Tube (CRT) programmer allowed the user to type in the control program using familiar relay symbols. It offered an alternative to the tedious process of inserting the program by using a manual loading process that varied in complexity depending on the machine. With the CRT display, the control logic could be viewed in the same form as on relay drawings, thus aiding the trouble-shooting process.

The addition of arithmetic functions and improved instruction sets expanded the application of programmable controllers by allowing them to be used with instrumentation devices that provided numerical input data. Logic and sequencing tasks could now be enhanced with the ability to perform calculations based on measured data. That same data could be used as the basis for corrective action. This newly found intelligence was the beginning of many advancements to come.

### Later Innovations

Hardware and software enhancements between 1975 and 1979 added even greater flexibility to the programmable controller. Improvements included larger memory capacity, remote input/output, analog and positioning control, operator communications, and software enhancement. These advancements made the programmable controller suitable for an even wider range of applications and contributed greatly to the reduction of wiring and installation cost.

Expanded memory systems allowed storage of larger application programs and amounts of data. Added memory allowed control programs to include not only logic and sequencing, but data acquisition and manipulation. The ability to store more data meant that prestored control data or recipes could be stored and retrieved automatically. For example, if a certain event occurred, all timer preset values could be changed, or high limit presets altered. This flexibility eliminated the need for operators to stop the process to change parameters.

Wiring costs were significantly reduced with the new ability to locate input/output subsystems in remote locations away from the Central Processing Unit (CPU) and nearby the controlled equipment. Instead of bringing hundreds of wires back to the CPU, the signals from the subsystems could be multiplexed over two twisted pairs of wire. Remote input/output systems also allowed large systems to be divided into smaller subsystems, which greatly improved maintainability and allowed a gradual start-up of major subsystems.

With the development of analog control, the programmable controller bridged the gap between on/off control systems and instrumentation controls. Up until then, the programmable controller, being capable of on/off control only, had been limited to partial control of applications such as chemical batching, water and waste treatment, and mineral processing. Applications such as these required a combination of on/off and variable (analog) control functions.

Another hardware development during this period were provisions to perform positioning control, using stepper output and encoder input feedback. The input interface counts a train of incoming pulses which provides a numeric value to the controller for verification of a move. Using data sent to it by the controller, the output interface produces a pulse train to be interpreted by the stepper motor translator. Early application of these interfaces included grinders, transfer heads, and paint spray lines.

Enhanced communication allowed PCs to communicate with other devices that helped to improve operator interface. The CRT hardware and software were further developed to aid program entry and monitoring. Production summaries, management reports, and maintenance data could be provided on hard copy through a printer. In late 1979, high-speed communication networks (data highways) evolved. These local networks allowed the control tasks of an entire plant to be distributed among several controllers, all communicating as one. With machines and processes communicating to one another, human communications could be eliminated from the control scheme. The data highway opened new applications to the programmable controller, such as machine transfer lines and material handling and tracking.

Software enhancements brought about computer-like statement instructions. These new instructions allowed easy utilization of the many hardware enhancements. Improvements included statements for manipulating and handling large amounts of data, as well as others for communicating with analog and peripheral devices. The relay-type instructions would have rendered these tasks very cumbersome or even impossible. Other software enhancements included system routines to improve on-line monitoring of the process. Menu driven routines simplified operator interface merely to selecting a numbered function.

With the developments during this period, the programmable controller took the first step toward replacing the minicomputer in many industrial applications.

## Today's Controllers

The early 1980s brought about many technological advances in the programmable controller industry. This explosion of capabilities reflected remarkable achievements in applications of microprocessor technology and helped fiercely competitive manufacturers who were seeking to attract and maintain customers. These changes not only affected the controller design, but also the philosophical approach to control system design. The following list describes some of the enhancements.

Hardware Enhancements:

- Faster scan times were achieved using bit-slice technology.
- Small low cost PCs (see Fig. 1-1a) replaced from 4-10 relays and also reduced space requirements.
- High density I/O systems (see Fig. 1-1b) provided space-efficient interfaces at low cost.
- Intelligent I/O (microprocessor based) interfaces expanded distributed processing. Typical interfaces are PID (proportional, integral, and derivative control), ASCII communication, positioning, host computer, and language modules (e.g. BASIC).
- Special interfaces that allow certain devices to be connected directly to the controller. Typical interfaces include thermocoupler, strain gauge, and fast response inputs.
- Mechanical design improvements included rugged I/O incasements and I/O systems that made the terminal an integral unit.
- Peripheral equipment improved operator interface and system documentation.

A significant hardware enhancement was the development of programmable controller families like that shown in Fig. 1-1c. These families consisted of a product line that ranged from very small single board *microcontrollers,* with as few as 10 I/O, to very large sophisticated PCs, with as many as 8000 I/O points and 128,000 words of memory. These family members utilizing common I/O systems and programming peripherals, typically could be interfaced to a local communications network. The family concept was an important cost savings development for users.

Software Enhancements:

- High-level languages, such as BASIC were implemented in some controllers to provide greater programming flexibility when communicating with peripheral devices. Hybrid type high-level languages were also implemented for control programming.
- Functional block instructions were implemented for ladder diagram instruction sets to provide enhanced software capability, but with simple program preparation.
- Diagnostics and fault detection were expanded from *system diagnostics,* which diagnose controller malfunctions, to include *machine diagnostics,* which diagnose failures or malfunctions of the controlled machine or process.
- Floating point math made it possible to perform complex calculations for control applications that required gauging, balancing, or statistical computation.
- Data handling and manipulation instructions were improved to accommodate complex control and data acquisition applications that involve storage, tracking, and retrieval of large amounts of data.

The programmable controller is now a mature control system offering much more than was ever anticipated. It is now capable of communicating with other control systems, providing production reports, scheduling production, and diagnosing its own failures and those of the machine or process. These enhancements have made the programmable controller an important contributor in meet-

**Figure 1-1.**

**(a)** PLC-Microtrol, 32 I/0 and 640 words of memory. (Allen-Bradley, Systems Div.)

**(b)** High density of I/O rack for use with the PC-700 and PC-900. 128 discrete I/O in a space of 19 by 12¼ inches. (Westinghouse Numa-logic Div.)

**(c)** Texas-Instruments PC Family: model 510 (center foreground), model 500 I/O system (rt. background), model 520 (lt. background), model 530 (cntr. background).

ing today's demands of higher quality and productivity. In spite of the fact that the programmable controller has become much more capable, it still has the simplicity of operation that was originally intended.

## Meeting The Future

Looking into the future, we can expect the most important changes to be continuations of current trends. New developments in hardware will include features such as advanced nonvolatile random access memory with error-correcting ability; user oriented operator interfaces; universal programming devices; communication between different brands of controllers; advanced fiber optic communication links; and faster, more flexible scan times. Software developments will include features such as multi-lingual controllers, canned control and data acquisition/reporting software packages, and user oriented languages with english-like statements.

The control philosophy of the future will cast programmable controllers in an important role in achieving factory automation. Control strategies will be distributed instead of centralized. PCs will be integrated with other control equipment such as robots, numerical controls, CAD/CAM systems, hierarchal PCs, and management information systems to achieve higher levels of quality and productivity in the factory of the future.

## 1-2　PRINCIPLES OF OPERATION

A programmable controller is a solid-state device used to control machine or process operation by means of a stored program and feedback from input/output devices. The National Electrical Manufacturers Association (NEMA) defines a programmable controller as a *digital electronic apparatus with a programmable memory for storing instructions to implement specific functions, such as logic, sequencing, timing, counting, and arithmetic, to control machines and processes.*

A programmable controller is composed primarily of two basic sections: the *Central Processing Unit (CPU)* and *Input/Output (I/O) Interface.* These sections are illustrated in Fig. 1-2.

Figure 1-3 is a closer look at the CPU. There are three main parts: the *Processor, Memory,* and *Power Supply.* Together, these components provide the intelligence of the controller. The CPU accepts (reads) input data from various sensing devices, executes the stored user program from memory, and sends appropriate output commands to control devices. This process of reading inputs, executing the program, and controlling outputs is done on a continuous basis and is called *scanning.* The power supply provides all the necessary voltages required for the proper operation of the other CPU sections.

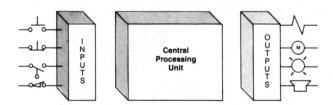

**Figure 1-2.** Programmable controller — block diagram.

8

CPU

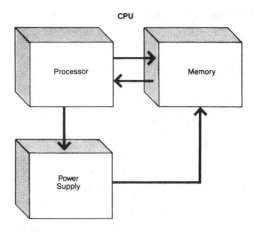

**Figure 1-3.** Block diagram of major CPU components.

The Input/Output system forms the interface by which field devices are connected to the controller. The purpose of the interface is to condition the various signals received from or sent to external (field) devices. Incoming signals from sensors such as pushbuttons, limit switches, analog sensors, selector switches, and thumbwheel switches are wired to terminals on the input interfaces. Devices that will be controlled, like motor starters, solenoid valves, pilot lights, and position valves, are connected to the terminals of the output interfaces.

Another programmable controller component is shown in Fig. 1-4. Although not generally considered a part of the controller, the *programming device* is required to enter the control program into memory. The programming device must be connected to the controller only when entering or monitoring the program. A CRT is commonly used for program entry and display, but alternate methods are available.

In Chapter 4, we will present a more detailed discussion of the Central Processing Unit and how it interacts with memory and the Input/Output interface. Chapter 5, is dedicated to the discussion of the I/O system.

**Figure 1-4.** APC programmable controller system; CPU, I/O, CRT programming unit (Courtesy, Cincinnati Milacron).

## 1-3  USING PROGRAMMABLE CONTROLLERS

### PCs vs Relays

For years, the question many engineers, plant managers, and Original Equipment Manufacturers (OEM) had asked was, *"Should I be using a programmable controller?"* At one time, much of a systems engineer's time was spent trying to determine the cost effectiveness of PC or relay control. Even today, many panel builders and control systems designers still think that they are faced with this decision. One thing is certain: today's demand for high quality and productivity can hardly be fulfilled economically without electronic control equipment. With rapid developments and increasing competition, the cost for programmable control has been driven down to the point where the old PC versus relay study is no longer necessary or valid. Programmable controller application can now be evaluated on its own merits.

With that thought in mind, certain questions can be asked concerning control needs.

- Is there a need for flexibility in control logic changes?
- Is there a need for high reliability?
- Are space requirements minimal?
- Are increased capability and output required?
- Are there data collection requirements?
- Will there be frequent control logic changes?
- Will there be a need for rapid modification?
- Must similar control logic be used on different machines?
- Is there a need for future growth?
- Can plant personnel operate and maintain the PC system?
- What are the overall costs?

The merits of programmable control systems make them especially suitable for applications in which the requirements listed above are particularly important for the economic viability of machine or process operation. A case which speaks for itself shown in Fig. 1-5, shows why PCs are easily favored over relays.

If system requirements call for flexibility or future growth, a programmable controller brings returns that will outweigh any initial cost advantage that relay-control might have. Even in the case which no flexibility or future expansion is required, a large control system can benefit tremendously from the trouble-shooting and maintenance aids a PC provides. The extremely short cycle (scan) time of a PC would allow the productivity of machines, which were previously under electro-mechanical control, to increase considerably.

Again, although relay-control may initially cost less, this advantage would be lost, if production downtime due to failures is high. These and other considerations for selecting a programmable controller are discussed later in Chapter 13.

### PCs vs Computers

The architecture of the programmable controller is basically the same as of a general purpose computer. In fact, some of the early PCs were computer based. However, some important characteristics distinguish PCs from general purpose computers.

First, unlike computers, the PC was specifically designed to survive the not always stable conditions of the industrial environment. A well-designed PC can be placed in areas with substantial amounts of electrical noise, electro-magnetic interference, mechanical vibration, or extreme temperatures and non-condensing humidity.

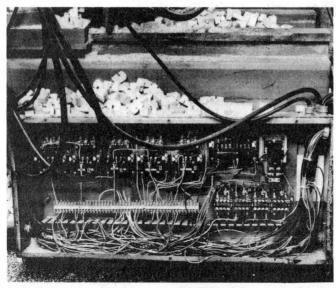

**Figure 1-5.** Illustration of relay control panel (top), converted to PC control panel (bottom) (Courtesy of Texas Instruments Corp.).

A second distinction of PCs is that the hardware and software are designed for easy use by plant electricians and technicians. The hardware interfaces for connecting the field devices are actually a part of the PC and are easily connected. The modular and self-diagnosing interface circuits pinpoint malfunctions and are easily removed and replaced. The software programming uses conventional relay ladder symbols or other easily learned languages.

The final distinction is one that is changing as PCs become more intelligent. Whereas computers are complex computing machines capable of executing several programs or tasks simultaneously and in any order, the PC executes a single program in an orderly and sequential fashion from first to last instruction. In recent years, however, flexibility has been added to the PC by including instructions that allow subroutine calling, interrupt routines, and a means for bypassing or jumping certain instructions.

## 1-4 TYPICAL AREAS OF APPLICATION

Since its invention, the programmable controller has been successfully applied in virtually every segment of industry including steel mills, paper and pulp plants, food processing plants, chemical and petrochemical plants, and automotive and power plants. PCs perform a great variety of control tasks, from repetitive on/off control of a simple machine to sophisticated manufacturing and process control. Table 1-1 lists a few of the major areas in which PCs have been applied and some of the typical applications.

**Table 1-1.** Typical Programmable Controller Applications

| CHEMICAL/PETROCHEMICAL | GLASS/FILM |
|---|---|
| Batch Process | Process |
| Materials Handling | Forming |
| Weighing | Finishing |
| Mixing | Packaging |
| Finished Product Handling | Palletizing |
| Water/Waste Treatment | Materials Handling |
| Pipeline Control | Lehr Control |
| Off-Shore Drilling | Cullet Weighing |
| | |
| MANUFACTURING/MACHINING | FOOD/BEVERAGE |
| Energy Demand | Bulk Materials Handling |
| Tracer Lathe | Brewing |
| Material Conveyors | Distilling |
| Assembly Machines | Blending |
| Test Stands | Container Handling |
| Milling | Packaging |
| Grinding | Filling |
| Boring | Weighing |
| Cranes | Finished Product Handling |
| Plating | Sorting Conveyors |
| Welding | Accumulating Conveyors |
| Painting | Load Forming |
| Injection/Blow Molding | Palletizing |
| Metal Casting | Warehouse Storage/Retrieval |
| Metal Forming | Loading/Unloading |
| | |
| MINING | METALS |
| Bulk Material Conveyors | Blast Furnace Control |
| Ore Processing | Continuous Casting |
| Loading/Unloading | Rolling Mills |
| Water/Waste Management | Soaking Pit |
| | |
| PULP/PAPER/LUMBER | POWER |
| Batch Digesters | Coal Handling |
| Chip Handling | Burner Control |
| Coating | Flue Control |
| Wrapping/Stamping | Load Shedding |
| Sorting Winding/Processing | |
| Woodworking | |
| Cut-to-Length | |

Although the uses of the programmable controller are far too extensive to enumerate, the applications shown here are frequently found within the listed industries. Later in Chapter 12, we describe several actual applications.

## 1-5 PRODUCT RANGES

Until approximately 1979, the choice in programmable controllers was limited to two types. The first type was small, relatively inexpensive, and not too versatile. These controllers were designed for the purpose of converting small electro-mechanical relay systems to solid-state control systems. Programming was rather tedious, and software functions were limited. The second type of controller was large, expensive, and designed to bridge the gap between small controllers and minicomputers. The functional capability spread between these two classes of controllers made the choice limited, and somewhat difficult. Buying the inexpensive system may have meant trying to live within its limitations, while purchasing the expensive controller could result in wasteful extravagance.

With the wide range of products available today, the chance of closely matching a product to the application is much better, but the task of selecting the best product is much more difficult. This situation is largely related to the fact that conventional classification of products, based on input/output and memory capacity, is no longer valid.

Typically, PCs are categorized into three major segments — small, medium, and large — each with distinct characteristic features. The standard features inherent to a particular segment were once a definite function of size (i.e. I/O, memory). As the market continues to grow and new capabilities are implemented, the three major segments have become less distinct. For example, a small PC may now have some features once found only in large PCs. Thus, I/O and memory specifications alone do not provide enough information for selecting a product.

Today's programmable controller product ranges can be graphically illustrated as shown in Fig. 1-6. This chart is not definitive, but for practical purposes is valid. Conventional segmentation would have placed boundaries at 128 I/O count for small PCs, 1024 I/O count for medium PCs, and 2048 I/O count for large PCs. However, market growth has led to the development of controllers with less than 32 I/O count, *micro-controllers* which fall into the small PC classification, and very large controllers with up to 8192 I/O count, which exceeds the conventional large PC segment. Four different size PCs are shown in Fig. 1-7.

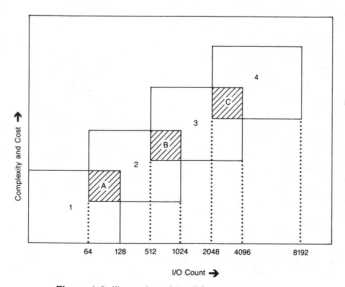

**Figure 1-6.** Illustration of the PC product range.

This new segmentation is divided into four major areas (1,2,3,and 4) with overlapping boundaries (A,B,and C). Area 1, small, includes controllers with capacity up to 128 I/O; area 2, medium, ranges up to 1024 I/O; area 3, large, with capacity up to 2048 I/O; and area 4, very large, defines controllers with a capacity up to 8192 I/O. The A, B, and C overlapping areas reflect enhancements of the standard features of PCs within a particular segment by adding options. These options allow a product to be closely matched to the application without having to purchase the next largest unit.

The major distinction among segments and similarities among overlapping areas will be covered in detail in Chapter 13. These differences are based on I/O count, memory size, programming language, software functions, and other factors. An understanding of the PC product ranges and their characteristics will allow the user to identify properly the controller that will satisfy a particular application.

**Figure 1-7.** Eptak 210 controller, top lt., 32 I/O (Eagle Signal Controls); Series Three, top rt., 400 I/O (General Electric); PLC-3, bottom lt., 8196 I/O (Allen-Bradley); model 584 M, bottom rt., 2000 I/O (Gould Modicon).

## 1-6  THE MANY BENEFITS

In general, PC architecture is modular and flexible, allowing hardware and software elements to expand as the application requirements change. In the event that an application outgrows the limitations of the PC, the unit can easily be replaced with a unit having greater memory and I/O capacity, and the old hardware can be reused for a smaller application.

### Flexible Control

Without question, the "programmable" feature provides the single greatest benefit from the installation of a programmable controller. Eliminating hardwired control in favor of programmable control is the first step towards achieving a flexible control system. Once installed, the control plan can be manually or automatically altered to meet the day-to-day control requirements, without changing the field wiring. This easy alteration is possible since there is no physical connection between the field input devices and output devices (see Fig. 1-8), as in a hardwired system. The only connection is through the control program, which can be easily altered.

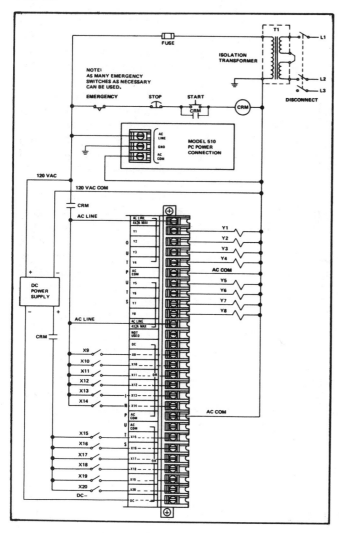

**Figure 1-8.** Illustration of typical PC connection — physical connection between inputs and outputs (Courtesy of Texas Instruments Corp.).

15

**Figure 1-9.** Installation the Micro 84 controller in a foundry sand compactability tester (Courtesy of Gould Modicon).

A typical example might involve a solenoid that is controlled by two limit switches connected in series. To change the solenoid operation by placing the two limit switches in parallel or combining a third switch to the existing circuitry could take less than one minute. In most cases, this simple program change can be made without shutting the system down. The same change to a hardwired system would have taken as much as thirty to sixty minutes of downtime. Even a half hour of downtime could mean a costly loss of production. A similar situation exists if there is a need to change a timer preset value or some other constant. A software timer in a PC can be changed in as little as five seconds. A set of thumbwheel switches and a pushbutton could be configured to input new preset values to any number of software timers very easily. The time-savings benefit of altering software timers as opposed to altering several hardware timers is obvious.

Similar flexibility and cost savings are provided by the hardware features of PCs. The intelligent CPU is capable of communicating with other intelligent devices. This capability allows the controller to be integrated into local or plantwide control schemes. With such a control configuration, the PC can send useful English messages regarding the controlled system to an intelligent display. On the other hand, the PC can receive supervisory information, such as production changes or scheduling information, from a host computer. The standard input/output system could include a variety of digital, analog, and other special interface modules that allow sophisticated control without using expensive, customized interface electronics.

## Easy Installation

Several PC attributes make each installation an easy and cost effective project. Its relatively small size allows the PC to be located conveniently in usually less than half the space required by an equivalent relay control panel (see Fig. 1-9). On a small scale changeover from relays, the PC's small and modular construction allows it to be mounted near the relay enclosure and prewired to existing terminal strips. Actual changeover can be made quickly by simply connecting the input/output devices to the prewired terminal strips.

In large installations, remote input/output stations are placed at optimum locations. The remote station is connected to the CPU by a twisted pair of wires. This configuration results in a considerable reduction of material and labor cost that would have been associated with running multiple wires and conduits. The remote subsystem approach also means that various sections of a total system could be completely prewired by an OEM or PC vendor, prior to reaching the installation site. This approach will mean a considerable reduction of the time spent by an electrician during on-site installation.

## Maintenance and Trouble-shooting

From the beginning, programmable controllers have been designed with ease of maintenance in mind. With virtually all components being solid-state, maintenance is reduced to the replacement of a modular, plug-in type component. Fault detection circuits and diagnostic indicators, incorporated in each major component, can tell if the component is working properly, or if it is malfunctioning. With the aid of the programming device, any programmed logic can be viewed to see if inputs or outputs are on or off. Programmed instructions can also be written to annunciate certain failures.

These and several other attributes of the PC make it a valuable part of any control system. Once installed, its contribution will be quickly noticed, and payback will be readily realized. The potential benefits of the PC, like any intelligent device, will depend on the creativity with which it is applied. Table 1-2 lists some of the many benefits available to PC users.

**Table 1-2.** Typical Programmable Controller Features/Benefits

| INHERENT FEATURES | BENEFITS |
|---|---|
| Solid-State Components | • High Reliability |
| Programmable Memory | • Simplifies Changes<br>• Flexible Control |
| Small Size | • Minimal Space Requirements |
| Microprocessor Based | • Communication Capability<br>• Higher Level Of Performance<br>• Higher Quality Products<br>• Multi-functional Capability |
| Software Timers/Counters | • Eliminate Hardware<br>• Easily Changed Presets |
| Software Control Relays | • Reduce Hardware/Wiring Cost<br>• Reduce Space Requirements |
| Modular Architecture | • Installation Flexibility<br>• Easily Installed<br>• Hardware Purchase Minimized<br>• Expandability |
| Variety Of I/O Interface | • Controls Variety Of Devices<br>• Eliminates Customized Control |
| Remote I/O Stations | • Eliminate Long Wire/Conduit Run |
| Diagnostic Indicators | • Reduce Trouble-shooting Time<br>• Signal Proper Operation |
| Modular I/O Interface | • Neat Appearance of Control Panel<br>• Easily Maintained<br>• Easily Wired |
| Quick I/O Disconnects | • Service w/o Disturbing Wiring |
| All System Variables Stored In Memory | • Useful Management/Maintenance Data Can Be Output in Report Form |

17

Without tarrying, we hasten to say that the potential benefits that stem from applying the programmable controller in an industrial application are innumerable. The list of benefits shown here has implications greater than first suggested. The bottom line is that through programmable control the user will achieve high performance and reliability that will result in high quality at reduced cost.

# CHAPTER
# --]2[--

# NUMBER SYSTEMS AND CODES

*When you measure what you are speaking about and express it in numbers, you know something about it; but when you cannot measure it, when you cannot express it in numbers, your knowledge is of a meager and unsatisfactory kind.*
—William Thomson, Lord Kelvin

This chapter introduces the number systems and digital codes that are most often encountered in programmable controller applications. Binary, octal, and hexadecimal number systems are most frequently used during input/output address assignment and program preparation. Binary Coded Decimal and Gray codes used in input/output applications, are also described. The ASCII character set is defined.

## 2-1  INTRODUCTION

A familiarity with number systems will prove quite useful when working with programmable controllers or with most any digital computer. This is true since a basic requirement of these devices is to represent, store, and operate on numbers, even to perform the simplest of operations. In general, PCs work on binary numbers in one form or another to represent various codes or quantities. Although number operations are transparent for the most part, and of little importance, there will be occasion to utilize a knowledge of number systems.

First let's review some basics. The following statements will apply to any number system.

- Each system has a base or radix.
- Each system can be used for counting.
- Each system can be used to represent quantities or codes.
- Each system has a set of symbols.

The base of a number system determines the total number of unique (different) symbols used by that system. The largest-valued symbol always has a value of one less than the base. Since the base defines the number of symbols, it is possible to have a number system of any base. However, number systems are typically chosen for their convenience. The number systems usually encountered while using programmable controllers are base 10, base 2, base 8, and base 16. These systems are also labeled *decimal, binary, octal,* and *hexadecimal* respectively. To demonstrate the common characteristics of number systems, let's turn to the familiar decimal system.

## 2-2  NUMBER SYSTEMS

### Decimal Number System

The decimal system, which is most common to us, was undoubtedly adopted as a result of man having ten fingers and ten toes. The base of the decimal number system is 10. The symbols or digits are 0,1,2,3,4,5,6,7,8, and 9. As noted earlier, the total number of symbols is the same as the base, and the largest-valued symbol is one less than the base. Because the decimal system is so commonly used, rarely do we stop to think how we express a number greater than 9. It is, however, important to note that the technique of representing a value greater than the largest symbol is the same for any system.

To express numbers greater than nine, a place value (weight) is assigned to each position that a digit would hold from right to left. The first position, starting from the right-most position, is 0; the second position is 1; and so on, up to the last position (n). The weighted value of each position can be expressed as the base (10 in this case) raised to the power of (n), the position. For the decimal system then, the position weights are 1,10,100,1000, etc.

Let's take for example, the number 9876:

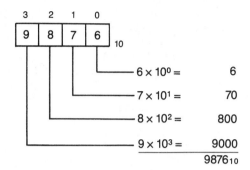

The value of the decimal number is computed by multiplying each digit by the weight of its position and summing the results. As we will see in other number systems, the decimal equivalent of any number can be computed by multiplying the digit times its BASE raised to the power of the digit's position. This is shown below.

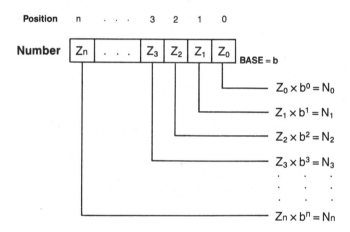

The sum of $N_0$ through $N_n$ will be the decimal equivalent of the number in base "b."

## Binary Number System

The binary numbering system uses the number 2 as the base. The only allowable digits are 0 and 1. There are no 2s, 3s, etc. The binary system is most accommodating for devices such as programmable controllers and digital computers. It was adopted for convenience, since it would be easier to design machines that distinguish between only two entities or numbers rather than ten as in decimal. Most physical elements have two states only: a light bulb is on or off, a valve is open or closed, a switch is on or off, a door is open or closed, and so on. With digital circuits, it is possible to distinguish between two voltage levels (e.g. +5V, 0V), which makes binary very applicable.

As with the decimal system expressing numbers greater than the largest-valued symbol (1) is accomplished by assigning a weighted value to each position from right to left. The decimal equivalent of a binary number is computed in the same way as for a decimal number, only instead of being 10 raised to the power of the position, it is 2 raised to the power of the position. For binary then, the weighted positions are 1,2,4,8,16,32,64, etc. For example:

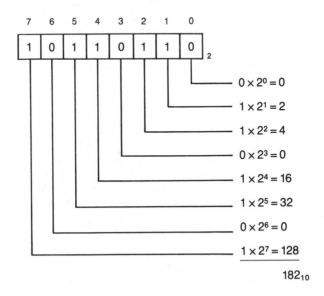

Since the binary number system uses only two digits, each position of a binary number can go only through two changes, and then a 1 is carried to the immediate-left position. Table 2-1 shows how binary numbers are used to count up to a decimal value of 10.

**Table 2-1.** Binary and Decimal counting

| Decimal | Binary |
|---------|--------|
| 0 | 0 |
| 1 | 1 |
| 2 | 10 |
| 3 | 11 |
| 4 | 100 |
| 5 | 101 |
| 6 | 110 |
| 7 | 111 |
| 8 | 1000 |
| 9 | 1001 |
| 10 | 1010 |

Each digit of a binary number is known as a "bit," therefore this particular binary number 10110110 (128 decimal) has 8 bits (or positions).

A group of 4 bits is known as a *nibble,* usually a group of 8 bits is a *byte,* and a group of one or more bytes is a *word.* Figure 2-1 shows a binary number composed of 16 bits, with the least and most significant bits.

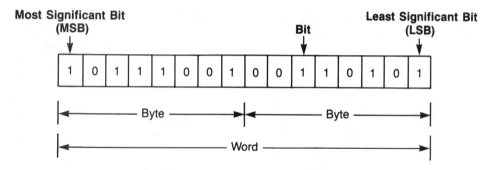

**Figure 2-1.** One word, two bytes, sixteen bits.

## Octal Number System

To express a number in binary requires substantially more digits than in the decimal system. For example, $91_{10} = 1011011_2$. Too many binary digits can become cumbersome to read or write when using large numbers, especially for human readers or writers. For convenience, the octal numbering system is brought into use. This system uses the number 8 as its base. The eight digits are 0,1,2,3,4,5,6, and 7. Like all other number systems, each digit in an octal number has a weighted decimal value according to its position. For example:

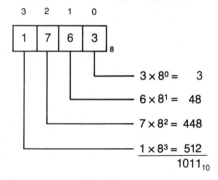

$$3 \times 8^0 = 3$$
$$6 \times 8^1 = 48$$
$$7 \times 8^2 = 448$$
$$1 \times 8^3 = 512$$
$$1011_{10}$$

As noted, octal is used as a convenient means of writing or handling a binary number. If a binary number is very large (many 1s and 0s), it can be represented by an equivalent octal number. As shown in the Table 2-2, one octal digit can be used to express three binary digits.

**Table 2-2.** Binary and related Octal Code

| Binary | Octal |
|--------|-------|
| 000 | 0 |
| 001 | 1 |
| 010 | 2 |
| 011 | 3 |
| 100 | 4 |
| 101 | 5 |
| 110 | 6 |
| 111 | 7 |

Since the base of the system is 8, or 2 to the 3rd power, any binary number can be represented as octal by grouping binary bits in groups of three. In this manner, a very large binary number can be easily represented by an octal number with significantly fewer digits. For example:

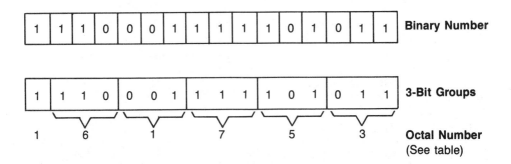

So, the 16 bit binary number can be represented directly by 6 octal digits.

As we will see later, most programmable controllers use the octal number system for referencing input/output and memory addresses.

## Hexadecimal Number System

The hexadecimal (hex) numbering system uses 16 as the base. It consists of 16 digits, numbers 0 thru 9, and the letters A thru F, which are substituted for the numbers 10 to 15 respectively. These 16 hexadecimal numbers are shown in Table 2-3, along with their decimal and binary equivalents.

**Table 2-3.** Hexadecimal code related to binary and decimal

| Hexadecimal | Binary | Decimal |
|-------------|--------|---------|
| 0 | 0000 | 0 |
| 1 | 0001 | 1 |
| 2 | 0010 | 2 |
| 3 | 0011 | 3 |
| 4 | 0100 | 4 |
| 5 | 0101 | 5 |
| 6 | 0110 | 6 |
| 7 | 0111 | 7 |
| 8 | 1000 | 8 |
| 9 | 1001 | 9 |
| A | 1010 | 10 |
| B | 1011 | 11 |
| C | 1100 | 12 |
| D | 1101 | 13 |
| E | 1110 | 14 |
| F | 1111 | 15 |

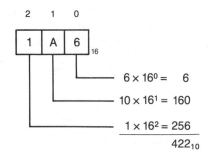

$$6 \times 16^0 = 6$$
$$10 \times 16^1 = 160$$
$$1 \times 16^2 = 256$$
$$422_{10}$$

As with the other number systems, hexadecimal numbers can be represented by their decimal equivalents by using the sum of the weights method. For example:

The value of A must be taken as 10 since A times 16 has to be expressed numerically, and A is equal to 10 decimal. The same rule applies thru F (10 thru 15). Like octal numbers, hexadecimal numbers can be easily converted to binary numbers without any mathematical conversion. Conversion is accomplished by writing the 4 bit binary equivalent of the hex digit of each position. For example:

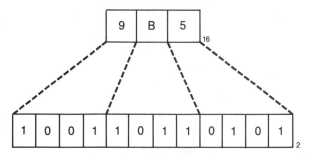

## 2-3 NUMBER CONVERSIONS

In the previous section we saw how a number of any base can be converted to decimal using the sum of the weights method. In this section, we show how a decimal number is converted to binary or octal.

To convert a decimal number to its binary equivalent, we must perform a series of divisions by 2 (system's base). The conversion process starts by dividing the decimal number by 2. If there is a remainder, it is placed in the least significant bit (LSB) of the binary number. If there is no remainder, a 0 is placed in the LSB. The remainder is brought down, and the process repeated until the result of the successive division has been reduced to 0. Conversion of the decimal number 37 to binary is shown here:

| Operation | | Remainder | |
|---|---|---|---|
| 2 | 37 | | |
| 2 | 18 | 1 | LSB |
| 2 | 9 | 0 | |
| 2 | 4 | 1 | |
| 2 | 2 | 0 | |
| 2 | 1 | 0 | |
| | 0 | 1 | MSB |

Thus, the binary equivalent of $37_{10}$ is $100101_2$. Another way, perhaps a faster method if dealing with large numbers, would be to divide continously by 8 in order to convert first to octal to get 3 bits at one time. This method is shown below.

| Operation | Remainder |
|---|---|
| 8 \| 37 | 5 LSD |
| 8 \| 4 | 4 MSD |
| 0 | |

So, the octal equivalent would be $45_8$, and the binary equivalent is $100101_2$. A look at Table 2-2 shows that octal 4 is 100 binary, and octal 5 is 101 binary.

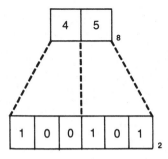

## 2-4  ONE'S AND TWO'S COMPLEMENT

One's and two's complement of a binary number is an operation used by programmable controllers, as well as computers, to perform internal mathematical calculations. To complement a binary number is to change it to a negative number. This allows the basic arithmetic operations of subtraction, multiplication, and division to be performed through successive addition. For example, to subtract the number 20 from 40 is done by first complementing 20 to obtain -20, and then performing an addition.

The intention here is to introduce the basic concepts of complementing rather than to provide a thorough analysis of arithmetic operations. However, several of the references listed in the back of this book cover this material.

### One's Complement

Let us assume that we have a 5 bit binary number that we wish to represent as a negative number. The number is decimal 23.

$$(10111)_2$$

There are two ways of representing this number as a negative number. The first method is simply to place the minus sign in front of the number, as with decimal numbers.

$$-(10111)_2$$

This method is suitable for us, but impossible for programmable controllers or computers, since the only symbols they use are binary 1s and 0s. To represent negative numbers then, digital computing devices use what is known as the complement method. First, the complement method places an extra bit (sign bit) in the most significant position (left-most), and lets this bit determine whether the number is positive or negative. The number is positive if the sign bit is 0 and negative if the sign bit is 1. Using the complement method, +23 would be represented as shown here:

$$0\ 10111_2$$

The negative representation of binary 23 using the 1s complement method, is simply obtained by inverting each bit, or changing 1s to 0s, and 0s to 1s. The one's complement of binary 23 then is shown here.

$$1\ 01000_2$$

If a negative number was given in binary, its complement would be obtained in the same fashion. Let's take the number -15:

$$1\ 0000 \longrightarrow -15$$
$$0\ 1111 \longrightarrow +15$$

## Two's Complement

Two's complement is similar to 1s complement in the sense that 1 extra digit is used to represent the sign. The two's complement computation, however, differs slightly. In 1s complement, all bits are inverted; but in two's complement, each bit from right to left is inverted, but only after the first "1" is detected. Let's use the number +22 as an example:

$$0\ 10110 \longrightarrow +22$$

Its two's complement would be:

$$1\ 01010 \longrightarrow -22$$

Note that in the positive representation of the number 22, starting from the right, the first digit is a 0 so it is not inverted; the second digit is a 1; all digits after this one are inverted.

If a negative number is given in two's complement, its complement (a positive number) is found in the same fashion.

$$1\ 10010 \longrightarrow -14$$
$$0\ 01110 \longrightarrow +14$$

All bits from right to left are inverted after the first 1 is detected. Other examples of two's complement are shown here:

$$0\ 10001 \longrightarrow +17 \qquad 0\ 00111 \longrightarrow +7$$
$$1\ 01111 \longrightarrow -17 \qquad 1\ 11001 \longrightarrow -7$$

$$0\ 00001 \longrightarrow +1$$
$$1\ 11111 \longrightarrow -1$$

The two's complement of 0 is not found, since no first 1 is ever encountered in the number. The two's complement of 0 then is still 0.

## 2-5   BINARY CODES

An important requirement of programmable controllers is to communicate with various external devices that supply information to the controller or receive information from the controller. This input/output function involves the transmission, manipulation, and storage of binary data that at some point must be interpreted by humans. Although the machine can easily handle this binary data, we require that the data be converted to a more interpretable form.

One way of satisfying this requirement is to assign a unique combination of 1s and 0s to each number, letter, or symbol that must be represented. This technique is called *binary coding.* In general, there are 2 categories of codes — those used to represent numbers only, and those that represent letters, symbols, and decimal numbers. Several codes for representing numbers, symbols, and letters have been instituted and are standard throughout the industry. Among the most common are the following:

- ASCII

- BCD

- GRAY

### ASCII

Alphanumeric (letters, symbols, decimal numbers) codes are used when information processing equipment, such as printers or CRTs, must handle the alphabet as well as numbers and special symbols. These characters — 26 letters (uppercase), 10 numerals (0-9), plus mathematical and punctuation symbols — can be represented using a 6 bit code (i.e. $2^6 = 64$). The most common code for alphanumeric representation is the *American Standard Code for Information Interchange* (ASCII). The acronym is pronounced "askey."

The ASCII code can be 6,7, or 8 bits. Even though the basic alphabet, numbers, and special symbols can be accommodated in a 6-bit code (64 possible characters), standard ASCII character sets use a 7-bit code ($2^7 = 128$ possible characters), which allows lower case and control characters for communication links, in addition to the characters already mentioned. This 7-bit code provides all possible combinations of characters used when communicating with peripherals or interfaces. The 8-bit ASCII code is used when parity check (see Chapter 4) is added to the standard 7-bit code for error checking. Note that all 8 bits can still be placed in one byte. Appendix E shows a standard ASCII table.

### BCD

*Binary Coded Decimal* was introduced as a convenient means for humans to handle numbers that need to be input to digital machines and to interpret numbers output from the machine. To handle numbers that otherwise would have been in binary would be extremely tedious, since we are more familiar with the decimal number system. The best solution to this problem is a means of converting a code readily handled by man (decimal) to a code readily handled by the equipment (binary). The result is BCD.

In decimal, we have the numbers 0 thru 9, whereas in BCD, each of these numbers is represented by a 4-bit binary number. Table 2-4 illustrates the relationship between the BCD code and the binary and decimal number systems.

**Table 2-4.** BCD code with binary and decimal equivalents

| Decimal | Binary | BCD |
|---------|--------|------|
| 0 | 0 | 0000 |
| 1 | 1 | 0001 |
| 2 | 10 | 0010 |
| 3 | 11 | 0011 |
| 4 | 100 | 0100 |
| 5 | 101 | 0101 |
| 6 | 110 | 0110 |
| 7 | 111 | 0111 |
| 8 | 1000 | 1000 |
| 9 | 1001 | 1001 |

The BCD representation of a decimal number is obtained simply by replacing each decimal digit by its BCD equivalent. The BCD representation of decimal 7493 is shown here as an example.

$$0111\ 0100\ 1001\ 0011$$
$$7\quad\ 4\quad\ 9\quad\ 3$$

Typical PC applications of BCD codes include data entry (time, volume, weights, etc.) via thumbwheel switches (TWS), data display via 7-segment displays, and input from absolute encoders. Because of the low cost of some integrated circuits, we find the necessary conversion of decimal to BCD and BCD to 7-segment already built into TWS and LED devices as illustrated in Fig. 2-2. This BCD data is then taken by the PC and converted internally into the binary equivalent of the input data through some instruction. To input or output BCD data, requires 4 lines to an input/output interface for each decimal digit.

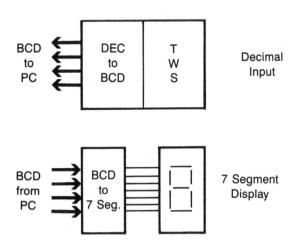

**Figure 2-2.** Illustration of thumbwheel switches (TWS), top; and seven segment LED display with converters.

## GRAY

The *Gray* code is one of a series of cyclic codes known as reflected codes and suited primarily for position transducers. It is basically a binary code that has been modified in such a way that only 1 bit changes as the counting number increases. In binary, as many as 4 digits could change when counting. This drastic change is seen in the transition from binary 7 to 8. Such a change allows a greater chance for error, which would be unsuitable for positioning applications. Most encoders use this code to determine angular position. Table 2-5 shows the Gray code and binary equivalent for comparison.

**Table 2-5.** Gray Code

| Gray code | Binary |
|-----------|--------|
| 0000 | 0000 |
| 0001 | 0001 |
| 0011 | 0010 |
| 0010 | 0011 |
| 0110 | 0100 |
| 0111 | 0101 |
| 0101 | 0110 |
| 0100 | 0111 |
| 1100 | 1000 |
| 1101 | 1001 |
| 1111 | 1010 |
| 1110 | 1011 |
| 1010 | 1100 |
| 1011 | 1101 |
| 1001 | 1110 |
| 1000 | 1111 |

In an optical absolute encoder, the rotor disk consists of an opaque and transparent segment, arranged in a Gray code pattern and illuminated by a light source that shines through the transparent sections of the rotating disk. The transmitted light is then received at the other end in Gray code form and is available for PC input in Gray code or BCD code if converted. Figure 2-3 illustrates a typical absolute encoder and its output.

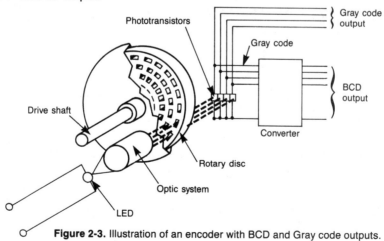

**Figure 2-3.** Illustration of an encoder with BCD and Gray code outputs.

## 2-6 REGISTER FORMATS

As mentioned earlier, a programmable controller performs all of its internal operations using binary 1s and 0s. Generally, these operations are performed using a group of 16 bits that represent numbers and codes. Recall that the grouping (unit) of bits on which a particular machine operates is called a word. A PC word is also called a *register* or *location*. Figure 2-4 illustrates a 16-bit register, comprised of 2 bytes.

Although the data stored in any register is represented in binary 1s and 0s, the format in which it is stored may differ from one controller to another. Generally, data is represented in straight (not coded) binary, or Binary Coded Decimal (BCD). Let's examine these two formats.

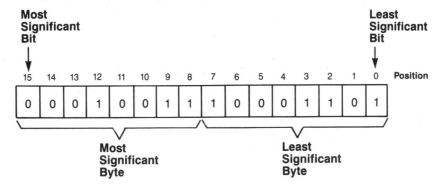

**Figure 2-4.** A 16-bit register.

### Binary Format

Data stored in this format can be directly converted to its decimal equivalent without any special restrictions. In a 16-bit register then, a maximum value of 65535 can be represented. Figure 2-5 shows the value 65535 in binary format (all bits are 1).

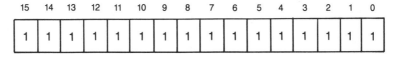

**Figure 2-5.** A register containing 65,535 in binary.

If the most significant bit of the register in Fig. 2-5 is used as a sign bit, then the maximum decimal value that can be stored is +32767, or —32767 (Fig. 2-6).

The decimal equivalent of these binary representations is achieved using the sum of the weights method. The negative representation of 32767 was derived using the two's complement method. As an exercise, see how these two numbers were computed (see previous section).

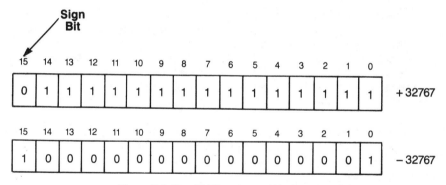

**Figure 2-6.** Two 16-bit registers with signed bit (MSB).

## BCD Format

If data is stored in a BCD format, then 4 bits are used to represent a single decimal digit. The only decimal numbers that these 4 bits can represent are 0 through 9. Our 16-bit register then can hold up to a 4 digit decimal equivalent, and the decimal values that can be represented are 0000-9999. The binary representation for BCD 9999 is shown in Fig. 2-7.

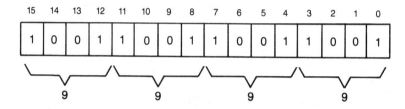

**Figure 2-7.** Register containing BCD 9999.

# CHAPTER

# --]3[--

# LOGIC CONCEPTS

*If it was so, it might be; and if it were so, it would be; but as it isn't, it ain't. That's logic.*
—Lewis Carroll

A fundamental prerequisite for understanding programmable controllers and their application is a working knowledge of logic operations. In this chapter and in later chapters, we will show how three basic logic functions — AND, OR, and NOT — can be combined to make very simple to complex control decisions.

## 3-1 THE BINARY CONCEPT

The binary concept is not a new idea; in fact, it is a very old one. It simply refers to the idea that many things can be thought of as existing in one of two states (bistable). For instance, a light can be ON or OFF, a switch OPEN or CLOSED, or a motor RUNNING or STOPPED. In digital systems, these two-state conditions can be thought of as a signal that is PRESENT or NOT PRESENT, ACTIVATED or NOT ACTIVATED, HIGH or LOW, ON or OFF, etc. This two-state concept can be the basis for making decisions, and since it is very adaptable to the binary number system, it is a fundamental building block for programmable controllers and digital computers.

Here as well as throughout this book, binary "1" represents the presence of a signal, or the occurrence of some event, while binary "0" represents the absence of the signal, or non-occurrence of the event. In digital systems, these two states are actually represented by two distinct voltage levels, as shown in Table 3-1. One voltage is more positive (or at a higher reference) than the other. Often, binary 1 (or logic 1) will be interchangeably referred to as TRUE, ON, or HIGH. Binary 0 (or logic 0) will be referred to as FALSE, OFF, or LOW.

**Table 3-1.** Binary Concept Using Positive Logic

| 1 (+5V) | 0 (0V) | EXAMPLE |
|---|---|---|
| OPERATED | NOT OPERATED | LIMIT SWITCH |
| RINGING | NOT RINGING | BELL |
| ON | OFF | LIGHT BULB |
| BLOWING | SILENT | HORN |
| RUNNING | STOPPED | MOTOR |
| ENGAGED | DISENGAGED | CLUTCH |
| CLOSED | OPEN | VALVE |

It should be noted in Table 3-1, that the more positive (HIGH) voltage, represented as logic 1, and the less positive (LOW) voltage, represented as logic 0, were arbitrarily chosen. The use of binary logic so that "1" represents the more positive voltage level or the occurrence of some event is referred to as *positive logic.*

*Negative logic,* as illustrated in Table 3-2, is the use of binary logic so that the more positive voltage level of the occurrence of the event is represented by "0"; "1" represented non-occurrence of the event or the less positive voltage level. Although positive logic is the more conventional of the two, negative logic sometimes may be more convenient.

## 3-2 AND, OR, and NOT FUNCTIONS

The binary concept showed how physical quantities (binary variables) that can exist in one of two states could be represented as "1" or "0." Now, we will see how statements that combine two or more of these binary variables can result in a true or false condition, represented by "1" and "0." The programmable controller will eventually make decisions based on the result of these statements.

**Table 3-2.** Binary Concept Using Negative Logic

| 1 (0V) | 0 (+5V) | EXAMPLE |
|--------|---------|---------|
| NOT OPERATED | OPERATED | LIMIT SWITCH |
| NOT RINGING | RINGING | BELL |
| OFF | ON | LIGHT BULB |
| SILENT | BLOWING | HORN |
| STOPPED | RUNNING | MOTOR |
| DISENGAGED | ENGAGED | CLUTCH |
| OPEN | CLOSED | VALVE |

Operations performed by digital equipment, such as programmable controllers, are based on three fundamental logic operations — AND, OR, and NOT. These operations are used to combine binary variables to form statements. Each function has a rule that will determine the statement outcome (true or false) and a symbol that represents the operation. For the purpose of this discussion, the result of a statement is called an *output (Y),* and the conditions of the statement are called *inputs (A,B).* Both the inputs and outputs represent two-state variables such as those discussed earlier in this section.

## The AND Function

The symbol shown in Fig. 3-1 is called an AND gate and is used to represent graphically the AND function.

Inputs ⊐⊃— Output

**Figure 3-1.** Symbol for AND function.

The AND output is True (1) only if all inputs are True (1).

The number of inputs to the AND gate is unlimited, but there is only one output. The truth table in Fig. 3-2 shows the resulting output (Y), based on all possible input combinations.

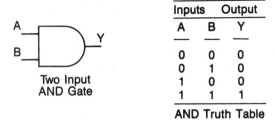

Two Input
AND Gate

| Inputs | | Output |
|--------|---|--------|
| A | B | Y |
| 0 | 0 | 0 |
| 0 | 1 | 0 |
| 1 | 0 | 0 |
| 1 | 1 | 1 |

AND Truth Table

**Figure 3-2.**

The following two examples show application of the AND function. A and B represent inputs to the controller.

**Example 3-1.** Basic Rule: If all inputs are "1," the output will be "1"; if any input is "0," the output will be "0."

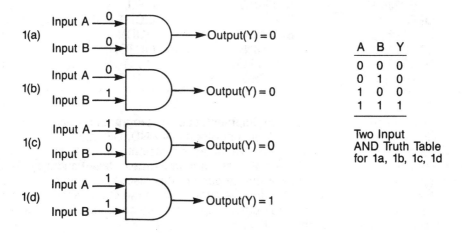

| A | B | Y |
|---|---|---|
| 0 | 0 | 0 |
| 0 | 1 | 0 |
| 1 | 0 | 0 |
| 1 | 1 | 1 |

Two Input
AND Truth Table
for 1a, 1b, 1c, 1d

**Example 3-2.** Light will be "ON" if switch A AND switch B are "Closed."

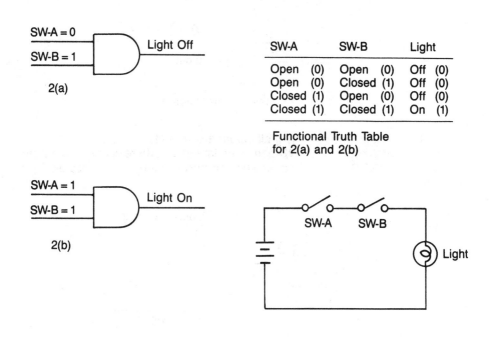

| SW-A | | SW-B | | Light | |
|---|---|---|---|---|---|
| Open | (0) | Open | (0) | Off | (0) |
| Open | (0) | Closed | (1) | Off | (0) |
| Closed | (1) | Open | (0) | Off | (0) |
| Closed | (1) | Closed | (1) | On | (1) |

Functional Truth Table
for 2(a) and 2(b)

## The OR Function

The symbol shown in Fig. 3-3 is called an OR gate and is used to represent graphically the OR function.

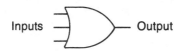

**Figure 3-3.** Symbol for OR function.

The OR output is True (1) if one or more inputs are True (1).

The number of inputs to an OR gate is unlimited, but there is only one output. The truth table in Fig. 3-4 shows the resulting output (Y), based on all possible input combinations.

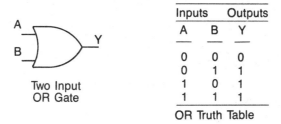

| Inputs | | Outputs |
|---|---|---|
| A | B | Y |
| 0 | 0 | 0 |
| 0 | 1 | 1 |
| 1 | 0 | 1 |
| 1 | 1 | 1 |

OR Truth Table

**Figure 3-4.**

The following examples show applications of the OR function. A and B are inputs to the controller.

**Example 3-3.** Basic Rule: If one or more inputs are "1," the output is "1"; if all inputs are "0," the output will be "0."

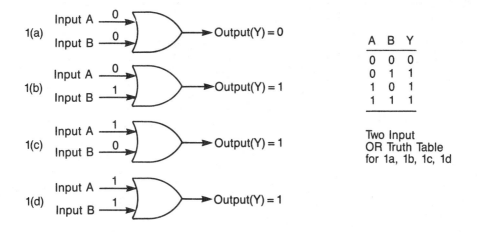

| A | B | Y |
|---|---|---|
| 0 | 0 | 0 |
| 0 | 1 | 1 |
| 1 | 0 | 1 |
| 1 | 1 | 1 |

Two Input
OR Truth Table
for 1a, 1b, 1c, 1d

**Example 3-4.** Light will be "ON" if switch A OR switch B is "Closed."

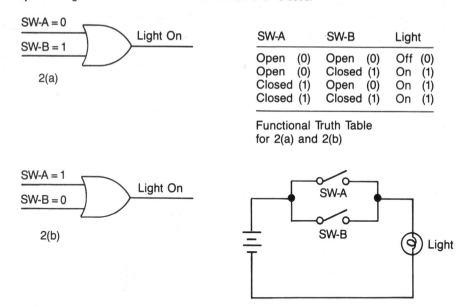

| SW-A | SW-B | Light |
|---|---|---|
| Open (0) | Open (0) | Off (0) |
| Open (0) | Closed (1) | On (1) |
| Closed (1) | Open (0) | On (1) |
| Closed (1) | Closed (1) | On (1) |

Functional Truth Table
for 2(a) and 2(b)

## The NOT Function

The symbol shown in Fig. 3-5 is the NOT symbol and is used to represent graphically the NOT function.

Input ──◯── Output

**Figure 3-5.** Symbol for NOT function.

The NOT output is True (1) if the input is False (0). The output is False (0) if the input is True (1). The result of the NOT operation is always the inverse of the input and is, therefore, sometimes called an *inverter*.

The NOT function, unlike the AND and OR, can have only one input and is never used alone, but in conjunction with the AND or the OR gate. The NOT operation and truth table are shown in Fig. 3-6.

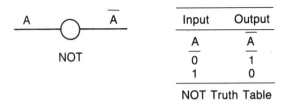

| Input | Output |
|---|---|
| A | $\overline{A}$ |
| 0 | 1 |
| 1 | 0 |

NOT Truth Table

**Figure 3-6.**

At first glance, application of the NOT function is not as easily visualized as the AND and OR functions. However, a closer examination of the NOT function shows it to be simple and quite useful. At this point, it would be helpful to recall three things that have been discussed.

1. Assigning 1 or 0 to a condition is arbitrary.
2. A "1" is normally associated with TRUE, HIGH, ON, etc.
3. A "0" is normally associated with FALSE, LOW, OFF, etc.

Examining 2 and 3 shows that logic 1 is normally expected to activate some device (e.g. if Y = 1, motor runs), and an output of logic 0 is normally expected to deactivate some device (e.g. if Y = 0, motor stops). If these conventions were to be reversed (Section 3-1), such that a logic 0 is expected to activate some device(e.g. if Y = 0, motor runs), and a logic 1 is expected to deactivate some device(e.g. Y = 1, motor stops), the NOT function would then have useful application.

1. A NOT is used when a "0" (Low condition) is to activate some device.
2. A NOT is used when a "1" (High condition) is to deactivate some device.

The following two examples show application of the NOT function. Although the NOT is normally used in conjunction with the AND and OR functions, for clarity the first example shows the NOT used alone.

**Example 3-5.** Basic Rule: If the input is "1," the output will be "0"; if the input is "0," the output will be "1."

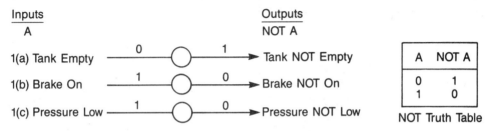

Inputs | Outputs
A | NOT A

1(a) Tank Empty — 0 — ◯ — 1 → Tank NOT Empty

1(b) Brake On — 1 — ◯ — 0 → Brake NOT On

1(c) Pressure Low — 1 — ◯ — 0 → Pressure NOT Low

| A | NOT A |
|---|-------|
| 0 | 1 |
| 1 | 0 |

NOT Truth Table

**Example 3-6.** Warning indicator. If power is ON AND pressure NOT okay (switch opened), the warning indicator will be ON.

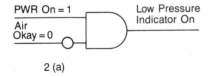

PWR On = 1
Air Okay = 0 — Low Pressure Indicator On

2 (a)

| Air Okay | PWR | Pressure Indicator |
|----------|-----|--------------------|
| 0 | 1 | 1 |
| 1 | 1 | 0 |

Truth Table for 2a and 2b.

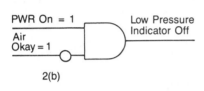

PWR On = 1
Air Okay = 1 — Low Pressure Indicator Off

2(b)

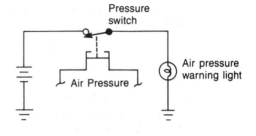

Pressure switch

Air Pressure

Air pressure warning light

Air pressure warning system

The two examples showed the NOT symbol placed on inputs to a gate. The NOT symbol placed at the output of an AND gate would negate or invert the normal output result. A negated AND gate is called a NAND gate. The logic symbol and truth table are shown in Fig. 3-7.

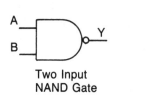

| Inputs | | Output |
|---|---|---|
| A | B | Y |
| 0 | 0 | 1 |
| 0 | 1 | 1 |
| 1 | 0 | 1 |
| 1 | 1 | 0 |

Two Input
NAND Gate

NAND Truth Table

**Figure 3-7.**

The same principle applies if a NOT symbol is placed at the output of the OR gate. The normal output is negated, and the function is referred to as a NOR. The symbol and truth table are shown in Fig. 3-8.

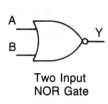

| Inputs | | Output |
|---|---|---|
| A | B | Y |
| 0 | 0 | 1 |
| 0 | 1 | 0 |
| 1 | 0 | 0 |
| 1 | 1 | 0 |

Two Input
NOR Gate

NOR Truth Table

**Figure 3-8.**

## 3-3  PRINCIPLES OF BOOLEAN ALGEBRA

An in-depth discussion of Boolean algebra is not required for the purpose of this book and is beyond the book's scope. However, an understanding of the technique of writing shorthand expressions for complex logical statements could be a useful tool when creating a control program of Boolean statements or conventional ladder diagrams.

Boolean algebra was developed in 1849 by an Englishman named George Boole. The intent of this algebra was to aid in the logic of reasoning, an ancient form of philosophy, rather than to implement the purposes of digital logic, a technique not developed at the time. It was meant to provide a simple way of writing complicated combinations of *logical statements*, defined earlier as statements that can be either TRUE or FALSE.

When digital logic finally came to be, Boolean algebra already existed as a simple way to analyze and express logic statements. All digital systems are based on this TRUE/FALSE or two-valued logic concept, where 1 represents TRUE and 0 represents FALSE. Because of this relationship between digital logic and Boolean logic, we will occasionally hear logic gates referred to as Boolean gates, or several interconnected gates called a Boolean network.

Figure 3-9 summarizes the basic operators of Boolean algebra, as they are related to the basic digital logic functions, AND, OR, and NOT. A capital letter represents the wire label of an input signal, a multiplication sign (•) represents the AND operation, an addition sign (+) represents the OR operation, and a bar over the letter ($\overline{A}$) represents the NOT operation. No other symbols are used in Boolean algebra.

| Logic Symbol | Logical Statement | Boolean Equation |
|---|---|---|
| A———Y (AND gate, inputs A, B) | Y is "1" if A and B are "1" | Y = A • B or Y = AB |
| C———Y (OR gate, inputs C, D) | Y is "1" if A or B is "1" | Y = C + D |
| A———Y (NOT gate) | Y is "1" if A is "0" Y is "0" if A is "1" | Y = $\overline{A}$ |

**Figure 3-9.** Summary of Boolean algebra as related to the AND, OR, and NOT functions.

In Fig. 3-9 the AND gate is shown with input signals A and B and an output signal Y. The output can be expressed by the logical statement: *Y is "1" if A AND B are "1."* The Boolean expression is written Y = A • B and is read *Y equals A ANDed with B*. The Boolean symbol (•) for AND could be removed and the expression written as Y = AB. Similarly, if Y is the result of ORing C and D, the Boolean expression is Y = C + D and is read *Y equals C ORed with D*. In the NOT operation, Y is the inverse of A, the Boolean expression is Y = $\overline{A}$ and is read *Y equals NOT A*. The basic Boolean operations of ANDing, ORing, and inversion are illustrated in Table 3-3. The table also illustrates how these three functions can be combined to obtain any desired logic combination.

**Table 3-3.** Logic Operations Using Boolean Algebra

**1. Basic Gates.** Basic logic gates implement simple logic functions. Each logic function is expressed in terms of a truth table and its Boolean expression.

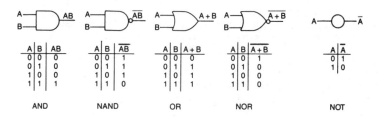

**2. Combined Gates.** Any combination of control functions can be expressed in Boolean terms using three simple operators: (•), (+), (—).

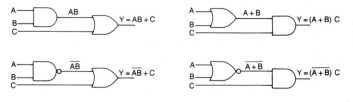

**Table 3-3.** Continued.

**3. Boolean Algebra Rules.** Control logic functions can vary from very simple to complex combinations of input variables. However simple or complex the functions may be, they satisfy these basic rules. The rules are a result of simple combination of the basic truth tables and may be applied to simplify logic circuits.

| | |
|---|---|
| $A + B = B + A$ | *COMMUTATIVE LAWS* |
| $AB = BA$ | |
| | |
| $A + (B + C) = (A + B) + C$ | *ASSOCIATIVE LAWS* |
| $A(BC) = (AB)C$ | |
| | |
| $A(B + C) = AB + AC$ | *DISTRIBUTIVE LAWS* |
| $A + BC = (A + B)(A + C)$ | |
| | |
| $A(A + B) = A + AB = A$ | *LAW OF ABSORPTION* |
| | |
| $\overline{(A + B)} = \overline{A}\,\overline{B}$ | *DE MORGAN'S LAWS* |
| $\overline{(AB)} = \overline{A} + \overline{B}$ | |

$$\overline{\overline{A}} = A, \overline{1} = 0, \overline{0} = 1$$

$$A + \overline{A}B = A + B$$

$$AB + AC + B\overline{C} = AC + B\overline{C}$$

**4. Order of Operations and Grouping Signs.** The order in which Boolean operations (AND,OR,NOT) of an expression is performed are important. This order will affect the resulting logic value of the expression. Consider the three input signals A,B,C. Combining them in the expression Y = A + B•C can result in misoperation of the output device(Y), depending on the order in which the operations are performed. Performing the OR operation prior to the AND operation is written (A + B)•C, and performing the AND operation prior to the OR is written A + (B•C). The result of these two expressions is not the same.

The order of priority in Boolean expressions is NOT (inversion) first, AND second, and OR last, unless otherwise indicated by grouping signs such as parentheses, brackets, braces, or the vinculum. According to these rules, the previous expression A + B•C, without any grouping signs, will always be evaluated only as A + (B•C). With the parentheses, it is obvious that B is ANDed with C prior to ORing the result with A. Knowing the order of evaluation then makes it possible to write the expression simply as A + BC, without fear of misoperation. As a matter of convention, the AND operator is usually omitted in Boolean expressions.

When working with Boolean logic expressions, misuse of grouping signs is a common occurrence. However, if signs occur in pairs, they do not generally cause problems if they have been properly placed according to the desired logic. Enclosing within parentheses two variables that are to be ANDed is not necessary, since the AND operator would normally be performed first. If, however, B is to be ORed with C prior to ANDing it with A, then B + C must be placed within parentheses.

When using grouping signs to insure proper order of evaluation of an expression, parentheses ( ) are used first. If additional signs are required, brackets , and finally braces [ ] are used. An illustration of the use of grouping signs is shown here.

$$Y1 = Y2 + Y5 \ [X1(X2 + X3)] + [Y4\,Y3 + X2(X5 + X6)]$$

**5. Application of DeMorgan's Laws.** DeMorgan's Laws are frequently used to simplify inverted logic expressions or simply to convert an expression into a useable form. See Appendix B for other conversions using DeMorgan's Laws.

According to DeMorgan's Laws:

$$\overline{AB} = \overline{A} + \overline{B} \qquad \text{and} \qquad \overline{A + B} = \overline{A}\,\overline{B}$$

## 3-4 HARDWIRED LOGIC TO PROGRAMMED LOGIC

Hardwired logic refers to logic control functions (timing, sequencing, and control) that are determined by the way devices are interconnected. In contrast to programmable control in which logic functions are programmable and easily changed, hardwired logic is fixed and changeable only by altering the way devices are connected. A prime function of the PC is to replace existing hardwired control logic or to implement control functions for new systems.

Relay logic implemented in PCs is based on the three basic logic functions (AND, OR, NOT) that were discussed in the previous sections. These functions are used either singly or in combinations to form instructions that will determine if a device is to be switched on or off. How these instructions are implemented to convey commands to the PC is called the language. The most widely used languages for implementing on/off control and sequencing are ladder diagrams, mnemonic statements, and Boolean equations. These languages are discussed at length in Chapter 7.

The most conventional of these languages is *ladder diagrams*. Ladder diagrams are also called *contact symbology,* since its instructions are relay-equivalent contact symbols (i.e. normally-open and normally-closed contacts and coils). Contact symbology is a very simple way of expressing the control logic in terms of symbols that are used on relay control schematics. If the controller language is ladder diagrams, the translation from existing relay logic to programmed logic is a one-step translation to contact symbology. If the language is mnemonic statements or Boolean equations, conversion to contact symbology is an unrequired step, yet still useful and quite often taken to provide an easily understood piece of documentation. Examples of translation from hardwired logic to programmed logic are shown later in Table 3-5.

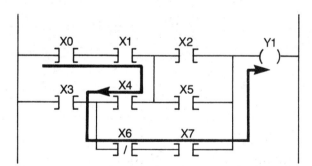

**Figure 3-10.** Typical ladder rung showing reverse power flow.

Figure 3-10 illustrates a typical ladder diagram rung. A rung is the contact symbology required to control an output. Some controllers allow a rung to have multiple outputs, but one output per rung is the convention. A complete ladder diagram program then consists of several rungs, each controlling an output. Each rung is a combination of input conditions (symbols) connected from left to right between two vertical lines, with the symbol that represents the output at the far right. The symbols that represent the inputs are connected in *series, parallel, or some combination* to obtain the desired logic. When completed, a ladder diagram control program consists of several rungs, with each rung having a specific output that it controls.

The programmed rung concept is a direct carryover from the hardwired relay ladder rung, in which input devices are connected in series and parallel to control various outputs. When activated, these input devices either allow current to flow through the circuit or cause a break in current flow, thereby switching a device on or off. The input symbols on a ladder rung can represent signals generated from connected input devices, connected output devices (see Table 3-4), or from outputs internal to the controller.

Each symbol on the rung will have a *reference number*, which is the *address* in memory where the current status (1 or 0) for the referenced input is stored. When the input signal is a connected input or output, the address is also related to the terminal where the signal wire is connected. The address then for a given input/output can be used throughout the program as many times as required by the control logic. This programmable controller feature is quite an advantage compared to relay-type hardware where additional contacts often mean additional hardware.

**Table 3-4.** ON/OFF input and output devices

| Input Devices | Output Devices |
|---|---|
| Pushbutton (PB) | Pilot Light (LT) |
| Selector Switch (SS) | Solenoid Valve (SOL) |
| Limit Switch (LS) | Horn (AH) |
| Proximity Switch (PRS) | Control Relay (CR) |
| Timer Contacts (TR) | Timer (TR) |

In Fig. 3-10, the inputs to the control logic are labeled X0 through X7, and the output is labeled Y1. How these symbols are normally referenced is dependent on the controller, but most are referenced using numeric addresses with octal (base 8) or decimal (base 10) numbering. Also illustrated is that any complete path (all contacts closed) from left to right will energize the output (Y1), with the exception of any path in which power flows in reverse. Power has to flow through X0, X1, down the vertical tie, then in reverse through X4, then through X6 and X7 to complete the path. Such a path is called a "sneak" path and is often required in hardwired logic. If the sneak path is required, adjustments to the ladder diagrams are easily made. In general, power flows from left to right, as well as up or down through vertical ties between parallel branches.

## Contact Symbols

The following symbols are used to translate relay control logic to contact symbolic logic. These symbols are also the basic instruction set for the Ladder Diagram, excluding timer/counter instructions. These and more advanced instructions are explained in Chapter 7.

Symbol                          Definition and Symbol Interpretation

—] [—          **Normally-opened contact.** Represents any input to the control logic. An input can be a connected switch closure or sensor, a contact from a connected output, or a contact from an internal output.

When interpreted, the referenced input or output is examined for an "ON" condition. If its status is "1," the contact will close and allow current to flow through the contact. If the status of the referenced input/output is "0," the contact will open and not allow current to flow through the contact.

—]/[— **Normally-closed contact.** Represents any input to the control logic. An input can be a connected switch closure or sensor, a contact from a connected output, or a contact from an internal output.

When interpreted, the referenced input/output is examined for an "OFF" condition. If its status is "0," the contact will remain closed, thus allowing current to flow through the contact. If the status of the referenced input/output is "1," the contact will open and not allow current to flow through the contact.

—( )— **Output.** Represents any output that is driven by some combination of input logic. An output can be a connected device or an internal output.

If any left-to-right path of input conditions is true (all contacts closed), the referenced output is energized (turned on).

—(/)— **NOT output.** Represents any output that is driven by some combination of input logic. An output can be a connected device or an internal output.

If any left-to-right path of input conditions are true (all contacts closed), the referenced output is de-energized (turned off).

The following seven points describe guidelines for translating from hardwired logic to programmed logic using contact symbols.

- **The normally-open contact.** When evaluated by the program, this symbol is examined for a "1" to close the contact; therefore, the signal referenced by the symbol must be ON, Closed, Activated, etc.

- **The normally-closed contact.** When evaluated by the program, this symbol is examined for a "0" to keep the contact closed; therefore, the signal referenced by the symbol must be Off, Open, De-activated, etc.

- **Outputs.** An output on a given rung will be energized if any left-to-right path has all contacts closed, with the exception of power flow going in reverse before continuing to the right. An output can control a connected device if the reference address is also a termination point, or internal output used exclusively within the program.

- **Inputs.** Contact symbols on a rung can represent input signals generated from connected inputs, contacts from internal outputs, or contacts from connected outputs.

- **Contact addresses.** Each program symbol is referenced by an address. If the symbol is referencing a connected input/output device, then the address is determined by the point at which the device is connected.

- **Logic format.** Contacts may be programmed in series or in parallel, depending on the logic required to control the output. The number of series contacts or parallel branches to a rung, is dependent on the controller.

- **Repeated use of contacts.** A given input, output, or internal output can be used throughout the program as many times as required.

The circuits in Table 3-5 show how simple hardwired series and parallel circuits can be translated to programmed logic. The series circuit is equivalent to the Boolean AND operation; therefore, all inputs must be ON to activate the output. The parallel circuit is equivalent to the Boolean OR operation; therefore, any one of the inputs must be ON to activate the output. Table 3-5 is explained in the following paragraphs.

**Table 3-5.** Example of translation from hardware logic to programmed logic.

| Relay Ladder Diagram | Contact or Ladder Diagram | Boolean Equation | Boolean Statements |
|---|---|---|---|
| a) Series Circuit | | $Y1 = X1 \cdot X2$ | STR  X1<br>AND  X2<br>OUT  Y1 |
| b) Parallel Circuit | | $Y2 = X3 + X4$ | STR  X3<br>OR  X4<br>OUT  Y2 |
| c) Series/Parallel Circuit | | $Y3 = (X5 + X6) \cdot C1$ | STR  X5<br>OR  X6<br>AND  C1<br>OUT  Y3 |
| d) Series/Parallel Circuit | | $Y4 = (X7 + X10) \cdot (C2 + C3)$ | STR  X7<br>OR  X10<br>STR  C2<br>OR  C3<br>OUT  Y4 |
| e) Parallel/Series Circuit | | $Y5 = (X11 \cdot X12) + X13$ | STR  X11<br>AND  X12<br>OR  X13<br>OUT  Y5 |
| f) Parallel/Series Circuit | | $C1 = (X14 \cdot X15) + (X16 \cdot X17)$ | STR  X14<br>AND  X15<br>OR  X16<br>AND  X17<br>OUT  C1 |
| g) Series Circuit | | $Y6 = X14 \cdot C1$ | STR  X14<br>AND  C1<br>OUT  Y6 |
| h) Series Circuit | | $Y7 = X14 \cdot \overline{C1}$ | STR  X14<br>AND<br>NOT  C1<br>OUT  Y7 |

**Series circuit (a).** In this circuit, if both switches LS1 AND LS2 are closed, the solenoid SOL1 will energize. According to this logic, a normally-opened contact symbols must be programmed.

**Parallel circuit (b).** In this circuit, if either of the two switches LS3 OR LS4 close, the solenoid SOL2 will energize. According to this logic, a normally-opened contact symbols must be programmed.

**Series/parallel circuits (c) and (d).** In a series/parallel circuit, the result of ORing two or more inputs is ANDed with one or more series or parallel inputs. In both of these examples, all of the relay circuit elements are normally-opened and must be closed to activate the pilot lights. Normally-open contacts are used in the program.

**Parallel/series circuits (e) and (f).** In a parallel/series circuit, the result of ANDing two or more inputs is ORed with one or more series inputs. In both these examples, all of the relay circuit elements are normally-opened, and must be closed to activate the output device. Normally-open contacts are used in the program.

**Internal outputs.** Circuit (f) controls an electromechanical control relay. Control relays do not normally drive output devices, but drive other relays. They are normally used to provide additional contacts for interlocking logic. The internal output provides the same function in software; however, the number of contacts are unlimited and can be either normally-open or normally-closed.

**Use of normally-open contacts.** Note in the series circuit (g) the solenoid will energize if LS14 closes and CR1-1 is energized. CR1-1 is a contact from the control relay CR1 in circuit (f) and closes whenever CR1 is energized. In the program, CR1 was replaced by the internal output C1; therefore, the program uses a normally-open contact from the internal output C1. SOL3 will energize when LS14 closes and C1 is energized.

**Use of normally-closed contacts.** In circuit (h), the solenoid will energize if LS14 closes, and CR1-1 is NOT energized. The program uses a normally-closed contact from the internal output C1. SOL3 will remain energized as long as the limit switch is closed, AND C1 is NOT energized.

# PART

# --]II[--

# UNDERSTANDING THE HARDWARE COMPONENTS

This part thoroughly explains the various hardware components. The reader should gain an in-depth understanding of the central processing unit and its relevance to the total control system, the input/output and peripheral interfaces, and their uses and connections.

# CHAPTER
# --]4[--

# THE CENTRAL PROCESSING UNIT

*Intelligence is the faculty of making artificial objects, especially tools, to make tools.*
—Henri Bergson

## 4-1  INTRODUCTION

The Central Processing Unit (CPU) is composed of three main sections: the processor, the system memory, and the system power supply. Figure 4-1 is a simplified illustration of the CPU. CPU architectures may differ from one manufacturer to another, but in general, most of them follow this typical organization. Although the power supply is shown inside the CPU enclosure, it may be a separate unit that is normally mounted next to the enclosure which contains the processor and memory. Figure 4-2 shows a CPU with a built-in power supply.

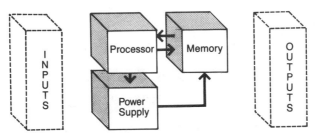

**Figure 4-1.** Block diagram of the CPU.

The term CPU is often used interchangeably with processor; in fact, however, the CPU term encompasses all the necessary elements that form the intelligence of the system. There are definite relationships between the sections that form the CPU and constant interaction among them. The processor is continually interacting with the system memory to interpret and execute the application program that controls the machine or process. The system power supply provides all the necessary voltage levels to insure proper operation of all the processor and memory components.

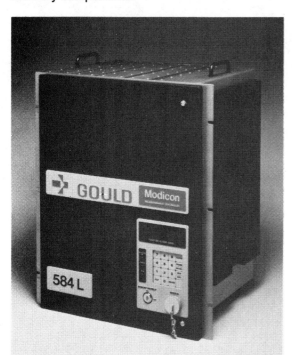

**Figure 4-2.** Model 584L high-speed programmable controller with built-in power supply (Courtesy of Gould-Modicon).

This chapter will cover the important aspects of the three main sections that form the CPU. An inside look at these sections will provide a better understanding of the functions and operations of a programmable controller. This understanding will be useful when specifying a product and defining the CPU requirements for your application.

## 4-2 PROCESSORS

The intelligence of today's programmable controllers is formed by very small *microprocessors* (micros), integrated circuits with tremendous computing and control capability. They perform all mathematical operations, data handling, and diagnostic routines that were not possible with relays or their predecessor, the hardwired processor.

The principal function of the processor is to command and govern the activities of the entire system. It performs this function by interpreting and executing a collection of system programs, known as the *executive*. The executive is a collection of supervisory programs that are permanently stored and considered a part of the controller itself. By executing the executive, the processor can perform all of its control, processing, communication, and other housekeeping functions. These programs allow communication with the processor via a programming device or other peripheral, monitoring of field devices, diagnosing of the system or the controlled machine or process, and executing the control program.

Several types of processor arrangements are used in PCs. Some controllers use a hardwired processor in conjunction with a micro to perform system duties.

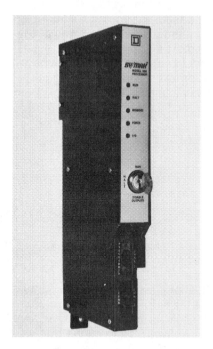

**Figure 4-3.** SY/MAX 300 processor module (Courtesy of Square D Co.).

Typically, with this arrangement, the hardwired processor executes the ladder program logic, while the microprocessor performs higher duties, such as the manipulation of data, math operations, and communications with other devices. Other controllers may use a single micro to perform all of the system functions. The processor components may reside on one printed circuit board or several. A processor module is shown in Fig. 4-3.

A newer approach is to divide the total system tasks among several micropro-cessors. This arrangement, in which a group of processors share the control and processing responsibility, is known as *multi-processing*. By allowing several pro-cessors to work together, the total system processing time is significantly reduced. Another multi-processor arrangement places the microprocessor intelligence away from the CPU. This technique involves intelligent *I/O interfaces* that contain a microprocessor, built-in memory, and a mini executive that performs independent control tasks. A typical intelligent module is the Proportional-Integral-Derivative (PID) control module, which performs closed loop control independent of the CPU.

Microprocessors used in PCs can also be categorized according to the word size that they use to perform operations. Standard word lengths are 4, 8, and 16 bits. This word length affects the speed in which most operations are performed. For example, a 16-bit microprocessor can manipulate data faster than an 8 bit micro, since it manipulates twice as much data in one operation. The difference in word length is of course, associated with the capability and degree of sophistication of the controller.

## Scan

During program execution, the processor reads all the inputs, takes these values, and according to the control logic, energizes or de-energizes the outputs, thus solving the ladder network. Once all the logic has been solved, the micro will update all outputs. The process of reading the inputs, executing the program, and updating the outputs is known as *scan*. Figure 4-4 illustrates a PC scan. The time required to make a single scan (scan time) can vary from 1 msec to 100 msec. PC manufacturers generally specify the scan time based only on the amount of application memory used (e.g. 10 msec/1K of programmed memory). However, the scan time is affected by other factors. The use of remote I/O subsystems increases the scan time as a result of having to transmit the I/O update to remote subsys-tems. Monitoring of the control program also adds overhead time to the scan, because the micro has to send the status of coils and contacts to the CRT or other monitoring device.

I/O Update

**Figure 4-4.** Illustration of a PC scan.

Program Scan

The scan is normally a continuous and sequential process of reading the status of inputs, evaluating the control logic, and updating the outputs. The common scan method of monitoring the inputs at the end of each scan is inadequate for reading certain extremely fast inputs. Some PCs provide software instructions that will allow the interruption of the continous program scan, in order to receive an input or update an output immediately. Figure 4-5 illustrates how immediate instructions operate during a normal program scan. These immediate instructions are very useful when the PC must react instantaneously to a critical input or output.

The scan time specification is an important consideration in the selection of a PC. It indicates how fast the controller can react to field inputs and correctly solve the control logic. For instance, if a controller has a total scan time of 10 msec and

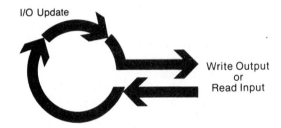

Program Scan

**Figure 4-5.** PC scan with immediate I/O update.

needs to monitor an input signal that can change states twice during an 8 msec period (less than the scan), the PC will never be able to "see" the signal, thus resulting in a possible machine or process malfunction.

## Subsystem Communications

This type of communication involves the information exchange that takes place between the CPU and an I/O subsystem. A subsystem can be located close to the CPU or at a remote location.

At the end of each program scan, the processor sends the latest status of outputs to the I/O subsystem and receives the current status of inputs. The actual communication between the processor and subsystem is performed via an I/O subsystem adapter module, located in the CPU, and a remote I/O processor module, located in the subsystem chassis or rack.

The distance between the CPU and a subsystem can vary depending on the controller and ranges between 1,000 and 15,000 feet. The communication media generally used are either a twisted pair or twinaxial, coaxial, or fiber optic cable, depending on the PC and the distance. The data transmission rate to subsystems is generally at very high speeds, but does vary depending on the controller. The data format also varies, but is normally a serial binary format of a fixed number of data bits (I/O status), start and stop bits, and error detection codes.

Error checking techniques are normally incorporated in the continuous communication between the processor and subsystem. These techniques are employed to confirm the validity of the data transmitted and received. The level of sophistication of error checking varies from one product to another, as does the type of errors reported and the resulting protective action. Let's look at some common error-checking methods.

## Error Checking

The processor uses error-checking techniques to monitor the functional status of the memory, communication links between subsystems and peripherals, and its own operation. Common error-checking techniques include *parity* and *checksum*.

**Parity.** Parity is perhaps the most common error detection technique, and it is used primarily in communication link applications to detect errors on long, error-prone data transmission lines. The communication between the CPU and subsystems is a prime example in which parity error-checking might be used. Parity check is sometimes called *Vertical Redundancy Check (VRC)*.

When using parity, the transmitted data is checked for an even or odd number of 1s. When data is transmitted, several 1s and 0s are sent. The number of 1s will either be odd or even, depending on the character or data being transmitted. In parity data transmission, an extra bit is added to the word, generally in the most significant or least significant bit position. This extra bit, called the *parity bit (P), is* used to make each byte or word have an odd or even number of 1s.

There are two types of parity checks: *even parity,* which checks for an even number of 1s, and *odd parity*, which checks for an odd number of 1s. If the transmission of data is to a subsystem, the parity type used will be already defined by the controller. However, if the data transmission is from the PC to a peripheral, the parity method used must be the same for both devices.

Suppose that the processor is to transmit the 7-bit ASCII character "C" (1000011) to a peripheral device, and odd parity is required. The total number of 1s is three, or odd. If the most significant bit is used for the parity bit (P), the transmitted data will be P1000011. To achieve odd parity, P is set to "0" to obtain an odd number of 1s. An error is detected at the receiving end if the data does not contain an odd number of 1s. If even parity had been the error-checking method, P would have been set to 1.

Parity error checking is a single-error detection method. If one bit of data in a word changes, an error will be detected due to the change in the bit pattern. However, if two bits change value, the number of 1s will be changed back, and an error will not be detected even though there is a mistransmission.

Some processors do not use parity when transmitting information, although it may be required by some peripherals. In this case, parity generation can be accomplished through application software. The parity bit can be set for odd or even parity, with a short routine using functional blocks or a high level language. If a non-parity oriented processor receives data that contain parity, a routine can also be used to mask-out or strip the parity bit.

**Checksum.** Many times, the extra bit of data added to each word, using parity error detection, is too wasteful to be desirable. In data storage, for example, error detection is desirable, but storing one extra bit for every 8 bits means a 12.5% loss of data storage capacity. For this reason, a data block error-checking method known as checksum is used.

Checksum error detection spots errors in blocks of many words instead of in individual words as parity does. This check is accomplished by taking all the words in a data block into consideration and adding to the end of the block one word that reflects a characteristic of the block. This last word, known as the *Block Check Character (BCC),* is shown in Fig. 4-6. This type of checking is appropriate for memory checks and is usually done at power up.

There are several methods of checksum computation. *Cyclic Redundancy Check (CRC)* and *Longitudinal Redundancy Check (LRC)* are two of the most common methods used. Other variations of checksum include the *Cyclic Exclusive-OR Checksum*.

The CRC performs an addition of all the words in the data block, and the resultant sum is stored in the last location (BCC). This summation process can rapidly reach an overflow condition. One variation of CRC will allow the sum to overflow and store only the remainder bits in the BCC word. Typically, the resulting word is complemented and written in the BCC location. During the error check, all words in the block are added together, and the addition of the final BCC word will turn the result to 0. A valid block can be detected by simply checking for 0 sum. Another type of CRC generates the BCC by taking the remainder after dividing the sum by a preset binary number.

The LRC is an error-checking technique based on the accumulation of the result of performing an Exclusive-OR of each of the words in the data block. The operation here is simply the logical Exclusive-OR of the first word with the second word, the result with the third word, and so on. The final Exclusive-OR operation is stored at the end of the block, as the BCC.

The Cyclic Exclusive-OR checksum is similar to LRC with some slight variations. The operation starts with a checksum word containing 0s, and an Exclusive-OR operation is done with the first word of the block. This is followed by a left rotate of the bits in the checksum word. The next word in the data block is Exclusive-ORed

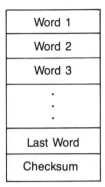

**Figure 4-6.** Illustration of data block with checksum.

with the checksum word and then rotated left. This procedure is repeated until the last word of the block is logically operated on. The checksum word is then appended to the block to become the BCC.

Most checksum error detection methods are usually performed by a software routine in the executive program. Typically, the processor performs the checksum computation on memory at power-up and also during the transmission of data. Some controllers, however, perform the checksum on memory during the execution of the control program. This continuous on-line error checking lessens the possibility of the processor operating on invalid data.

**Error Detection and Correction.** More sophisticated programmable controllers may have an error detection and correction scheme that provides greater reliability than conventional error detection. The key to this type of error correction is the multiple representation of the same value. If a single bit changes, the value would remain the same.

The most common error detection and correction code is the *Hamming Code*. This code relies on parity bits interspread with data bits in a data word. By combining the parity and data bits according to a strict set of parity equations, a small byte is generated that contains a value that actually points to the bit in error. An error can be detected and corrected if any bit is changed in any value. The hardware used to generate and check Hamming Codes is quite complex and essentially implements a set of error-correcting equations.

Error-correcting codes offer the advantage of being able to detect two or more bit errors; however, they can only correct one bit errors. They also present a disadvantage in that they are bit-wasteful. Nevertheless, this scheme will be seen in the future primarily in data communication in hierarchal systems that are completely unmanned, sophisticated, and automatic.

## CPU Diagnostics

The processor is responsible for detecting communication as well as other failures that may be encountered during system operation. It must alert the operator or system in case of a malfunction. To achieve this goal, the processor performs error checks during its operation and sends status information to indicators that are normally located on the front of the CPU. Typical diagnostics include memory OK, processor OK, battery OK, and power supply OK. Depending on the controller, a set of fault relay contacts may be available. The fault relay is controlled by the processor and is activated when one or more specific fault conditions (CPU diagnostics) occur. The fault relay contacts can be used in an alarm circuit to signal a failure.

The relay contacts that are generally provided with the controller operate in a "watch-dog" timer fashion. The processor sends a pulse at the end of each scan indicating a correct system operation. If there is a failure, the processor would not send a pulse, the timer would time-out, and the fault relay would activate.

In some controllers, certain CPU diagnostics are available to the user for use during the execution of the control program. These diagnostics use internal outputs that are controlled by the processor, but can be referenced by the user program (e.g. loss of scan, battery low, etc.).

## 4-3   MEMORY OVERVIEW

The term programmable controller implies that a sequence of instructions, or programs, and data must be stored somewhere. That somewhere is called the *memory system*. An understanding of what is stored in the programmable controller memory will aid in understanding why certain things are stored as they are and why certain considerations must be made when deciding on the memory type and capacity to best suit a particular application.

The programs and data stored in the memory system are generally described using the following four terms:

- *Executive:* A permanently stored collection of programs that are considered a part of the system itself. These supervisory type programs direct system activities such as execution of the control program, communication with peripheral devices, and other system housekeeping activities.

- *Scratch Pad:* A temporary storage used by the CPU to store a relatively small amount of data for interim calculations or control. Data that is needed quickly is stored in this area to avoid the access time that would be involved if it were stored in the main memory.

- *Applicaton Memory:* This area provides storage for any programmed instructions entered by the user. The control program is stored in this area.

- *Data Table:* This area, a part of the application memory, stores any data associated with the control program, such as timer/counter preset values, and any other stored constants or variables that are used by the control program or the CPU. It also retains the status information of the system inputs once they have been read and the system outputs once they have been set by the control program.

The storage and retrieval requirements are not the same for the executive, the scratch pad, the I/O buffer, the user program, and the data table; therefore, they are not always stored in the same types of memory. For example, the executive requires a memory that permanently stores its contents and cannot be deliberately or accidentally disturbed or altered by loss of electrical power or by the user. Such a memory for the application program or the data table may prove unsuitable.

Although there are several types, memory can be placed into the following two categories:

- *Volatile*
- *Nonvolatile*

Volatile memory will lose its programmed contents if all operating power is lost or removed, whether it is normal power or some form of back-up power. Volatile memory is easily altered and quite suitable for most applications, when supported by battery back-up and possibly a cassette recorded copy of the program.

Nonvolatile memory, will retain its programmed contents, given a complete loss of operating power. It requires no type of back-up. Nonvolatile memory generally is unalterable, yet there are special, nonvolatile memory types that are alterable.

Today's programmable controllers include those that use nonvolatile memory and volatile memory with battery back-up, as well as those that offer both. Volatile and nonvolatile memory types that are used in programmable controllers are explained in the following section.

## 4-4  MEMORY TYPES

There should be two major concerns regarding the type of memory where the application program is stored. Since this memory is responsible for retaining a control program that will be run each day, volatility    should be the prime consideration. Without the application program, production may be delayed or forfeited. The outcome is usually unpleasant. A second concern should be the ease with which the program stored in memory can be altered. Ease in altering the memory is important, since the memory is ultimately involved in any interaction that goes on between the user and the controller. This interaction begins with program entry and continues with program changes made during program generation, system start-up,   and ultimately on-line changes such as changing a timer or counter preset value.

The following discussion describes various types of memory, and how their characteristics affect the manner in which programmed instructions are retained or altered within the programmable controller.

### Read-Only Memory (ROM)

A Read-Only Memory is designed to store permanently a fixed program, which is not alterable under ordinary circumstances. It acquires its name from the fact that its contents can be examined or read, but not written into or altered once the data or program has been stored. ROM is contrasted to memory types that can be read from and written to (see RAM, Read/Write). Because of their nature, ROMs are generally immune to changes due to electrical noise or loss of power. Executive programs are often stored in a ROM.

**ROM as Application Memory.** Generally, programmable controllers rarely use ROM for their application memory. However, in applications that require fixed data, the Read Only Memory offers advantages when speed, cost, and reliability are factors. Generally, the creation of ROM-based PC programs is accomplished at the factory by the manufacturer. Once the original set of instructions is programmed, it can never be altered by the user. The typical approach to programming of ROM-based controllers assumes that the program has already been debugged. This would have to be accomplished using a Read/Write-based PC or possibly a computer. The final program can then be entered into ROM. ROM application memory is typically found only in very small, dedicated programmable controllers.

### Random Access Memory (RAM or R/W)

Random Access Memory, often referred to as *Read/Write Memory*, is designed so that information can be written into or read from any unique location. There are two types of RAM: the volatile RAM, which does not retain its contents if power is lost, and non-volatile RAM (see NOVRAM, Core), which retains its contents if power is lost. Volatile RAM normally has a battery backup to sustain it during power outages.

**RAM as Application Memory.** Today's controllers, for the most part use RAM with battery support for application memory. RAM provides an excellent means for easily creating and altering a program, as well as allowing data entry. In comparison to other memory types, RAM is a relatively fast memory. The only noticeable disadvantage of battery supported RAM is the fact that it requires a battery that may eventually fail. Battery supported RAM is sufficient for most applications. If a battery backup is not feasible, a controller with a nonvolatile memory option can be used in combination with the RAM (e.g. RAM, EPROM). This type of memory arrangement provides the advantages of both volatile and non-volatile memory. A RAM chip is shown in Fig. 4-7.

**Figure 4-7.** 4K by 8-bit RAM memory (4Kx1 bit/chip).

## Programmable Read Only Memory (PROM)

The PROM is a special type of ROM.

**PROM as Application Memory.** Less than 1% of today's programmable controllers use PROM for application memory. When it is used, this type of memory will most likely be a permanent storage backup to some type of RAM. Although a PROM is programmable and like any other ROM has the advantage of nonvolatility, it has the disadvantage of requiring special programming equipment, and once programmed it cannot be erased or altered. Any program change would require a new set of PROM chips. A PROM memory might be suitable for storing a program that has been thoroughly checked while residing in RAM and will not require further changes or on-line data entry.

## Erasable Programmable Read Only Memory (EPROM)

The EPROM is a specially designed PROM that can be reprogrammed after being entirely erased with the use of an *ultra-violet (UV) light source*. Complete erasure of the contents of each chip requires that the window of the chip (see Fig. 4-8) be exposed to the UV-light for approximately 20 minutes. The EPROM can be considered a semi-permanent storage device in that it permanently stores a program until it is ready to be altered. Program changes can be made after the EPROM chip is completely erased.

**EPROM as Application Memory.** EPROM provides an excellent storage medium for an application program in which nonvolatility is required, but program changes or on-line data entry is not required. Many Original Equipment Manufacturers (OEMs) use controllers with EPROM type memories to provide permanent storage of the machine program after it has been developed, debugged, and is fully operational. EPROM is probably used because most machines supplied by OEMs will not require any changes or data entry by the user.

An application memory composed of EPROM alone would be unsuitable, if on-line changes and/or data entries are a requirement. However, many controllers offer EPROM application memory as an optional backup to battery supported RAM. EPROM, with its permanent storage capability combined with the easily altered RAM, makes a suitable memory system.

**Figure 4-8.** 4K by 8-bit EPROM memory chip.

## Electrically Alterable Read Only Memory (EAROM)

Electrically Alterable ROMs are similar to EPROMs, but instead of requiring an ultraviolet light source to erase them, an erasing voltage on the proper pin of an EAROM chip would wipe it clean.

**EAROM as Application Memory.** Very few controllers use EAROM as application memory, but like EPROM, it provides a non-volatile means of program storage and can be used as a backup to RAM type memories.

## Electrically Erasable Programmable Read Only Memory (EEPROM)

EEPROM is a relatively new integrated circuit memory storage device, which was developed in the mid-1970s. Like ROMs or EPROMs, it is a non-volatile memory, yet it offers the same programming flexibility as does RAM.

**EEPROM as Application Memory.** Several of today's small- and medium-sized controllers are implementing EEPROM as the only memory within the system. It provides permanent storage of the program and can be easily changed with the use of the standard CRT or manual programming unit. These two advantages will certainly help to eliminate downtime, or delays associated with programming changes, and will also help to lessen two disadvantages of the EEPROM.

One disadvantage of EEPROM is that writing to a byte of memory is possible only after erasing that byte. The erase/write process takes approximately 10 and 15 milliseconds. This delay period is noticeable when an on-line program change is being made. Another characteristic of the present generation EEPROM is a limitation on the number of times that a single byte of memory can undergo the erase/write operation (approx. 10,000). These disadvantages are negligible when compared to the remarkable advantages that EEPROM offers.

## Core

Core memory is a nonvolatile type memory. It received its name from the fact that it stores individual bits by magnetizing small toroidal ferrite cores in the 1 or 0 direction, through a write-current pulse (Fig. 4-9). Each core represents a bit and can hold a 1 or 0 even in the event of power loss. Core memory units are nonvolatile because power is not necessary to keep the cores magnetized. The state of each core is also electrically alterable, which makes it a nonvolatile RAM.

**Core as Application Memory.** Core was used in many of the first programmable controllers and is still used in a few controllers today. It provides an excellent means for permanent storage and is easily altered. Distinct disadvantages of using core memory in programmable controllers are its slow speed, relatively expensive cost, and somewhat large physical space requirements.

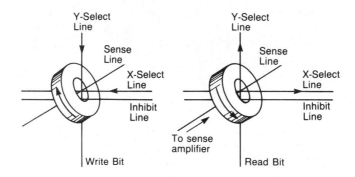

**Figure 4-9.** Illustration of a magnetic toroidal used in core memories.

## Non-Volatile Random Access Memory (NOVRAM)

NOVRAM, a newer memory fabrication, employs both the conventional RAM and the nonvolatile EEPROM on a single chip. Each bit within the RAM section of the chip has a corresponding adjacent bit within the EEPROM (see Fig. 4-10). Nonvolatile data can be stored in the EEPROM, and at the same time, independent data can be written to or accessed from the RAM. Data can be transferred back and forth between the RAM and the EEPROM at any time.

**NOVRAM as Application Memory.** At present, very few (mostly small) programmable controllers use NOVRAM technology. A NOVRAM application memory is the perfect solution to application requirements of easy reprogrammability and nonvolatility. When entered, the control program is stored directly into the RAM and is also executed from the RAM. An exact replica of the executed program is dumped into the nonvolatile EEPROM, without any user intervention, when the STORE operation takes place. When power is removed from the controller, the RAM contents are lost, but are restored by the RECALL (from EEPROM to RAM) command on each power-up. With this memory arrangement, the convenience of easy changes and a permanent copy of the program are provided in a single memory system, as opposed to two separate memory systems (e.g. RAM, EPROM).

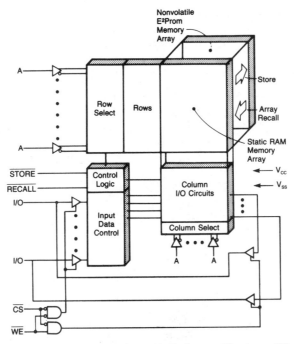

**Figure 4-10.** NOVRAM memory block diagram (Courtesy of Xicor).

NOVRAM can be considered the state of the art in application memory for programmable controllers, and will probably replace the current memory technologies used in PCs. Presently, however, because of its low density and relatively high cost, NOVRAM will probably be applied mostly in small controllers. Figure 4-11 shows an actual NOVRAM chip.

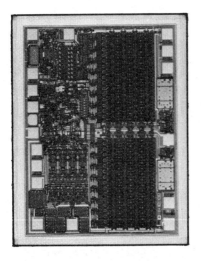

**Figure 4-11.** Die photo of NOVRAM 2210 memory circuit augmented (L) and actual chip (R) (Courtesy of Xicor).

## 4-5   MEMORY STRUCTURE AND CAPACITY

### Basic Structural Units (Bits, Bytes, Words)

Programmable controller memories can be visualized as a large two-dimensional array of single unit storage cells, each of which can store a single piece of information in the form of "1" or "0". It is obvious, then, that the binary numbering system is used to represent any information stored in memory. Since BIT is the acronym for BInary digiT, and each cell can store a bit, each cell is called a bit. A bit then is the smallest structural unit of memory and stores information in the form of 1s and 0s. Ones and zeroes are not actually in each cell; instead, each cell has a voltage charge present (indicating a 1) or not present (indicating a 0). The bit is considered ON if the stored information is 1, and OFF if the stored information is 0. The ON/OFF information stored in a single bit is referred to as *bit status*.

Sometimes it is necessary for the processor to handle more than a single bit. For example, it would be more efficient to handle a group of bits when transferring data to and from memory. To store numbers and codes, as we have discussed, also requires a grouping of bits. A group of bits handled simultaneously is a "byte". A byte is more accurately defined as the smallest group of bits that can be handled by the processor at one time. Although byte size is normally 8 bits, the size can vary depending on the specific controller.

The third and final structural information unit used within the programmable controller is called a "word". A word is also a fixed group of bits that varies according to the controller. In general, a word is the unit that the processor uses when data is to be operated on, or when instructions are to be performed. Word length is usually 1 byte in length or more. For example, a 16 bit word consists of 2 bytes. Typical word lengths used in PCs are 8, 12, and 16 bits. Figure 4-12 illustrates the structural units of a typical programmable controller memory.

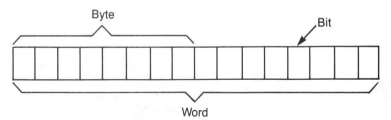

**Figure 4-12.** Basic structural units of PC memory.

### Memory Capacity

Memory capacity is a vital concern when considering a programmable controller application. Specifying the right amount of memory can mean a savings in cost of hardware as well as avoiding lost time, if additional memory capacity is required later. Knowing the memory capacity requirements ahead of time could also mean avoiding the possiblity of purchasing a controller that does not have adequate capacity or that is not expandable.

Memory capacity is expandable in some controllers, but not in others. The very small PCs that control 10 to 64 input/output devices, usually have fixed memory capacity. Controllers that handle 64 or more input/output devices are usually expandable in increments of 1K, 2K, 4K, etc. The abbreviation *(K)* representd 1024 locations when referring to memory capacity. A 1K memory contains 1024 storage locations, 2K is 2048 locations, 4K is 4096, and so on.

The memory capacity of a particular controller in units of (K) is only an indication of the total number of storage locations available. This maximum number alone is not enough when determining memory requirements. Additional information concerning how program instructions are stored will help in making a better decision. The term *memory utilization* refers to the amount of data that can be stored in one location or, more specifically, to the number of memory locations required to store each type of instruction. This data can be obtained from the manufacturer if not found in the product literature. Figure 4-13 illustrates the importance of knowing how instructions are stored before specifying memory requirements.

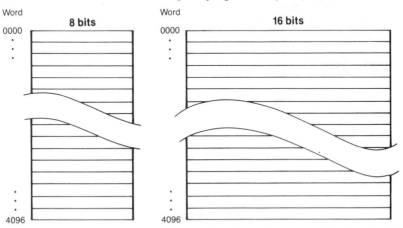

**Figure 4-13.** Block illustration of 4K (4096) storage locations using two specifications; (a) 4K by 8-bit locations and (b) 4K by 16-bits.

Suppose 16 bits were required to store each normally-open and normally-closed contact instruction. With this information, it becomes quite clear that the effective storage area of the memory system in Fig. 4-13a is one half that of Fig. 4-13b. This means that to store the same size control program would require 8K memory capacity instead of the specified 4K.

After becoming familiar with how memory is utilized in a particular controller, users can begin a first determination of the maximum memory requirements for the application. Although several rules of thumb have been used over the years, no one simple rule has emerged as being the most accurate. However, with a knowledge of the number of outputs, some idea of the number of program contacts needed to drive each output, and information concerning memory utilization, a first determination can be reduced to a simple multiplication.

**Example 4-1.** First determination of memory requirements

Control logic = 20 contacts/output

No. Outputs    = 50

Memory utilization = 1 word/contact or output (element)

Total contacts = 1000
Total outputs =    50

---

Outputs + Contacts = 1050

Memory required  = 1050 x 1 word/element

In this example, the memory requirement is 1050 word locations, or just over 1K. This calculation is considered a first determination, because other factors must be considered before the final decision is made. This exercise shows that the minimum memory requirement to program the 50 output sequences, is 1050 words at 20 contacts/output.

We should keep in mind that memory requirements are also affected by the sophistication of the control program. If the application requires data manipulation and data storage, additional memory will be required. Normally, the enhanced instructions that perform mathematical and data manipulation operations will have greater memory requirements. Exact usage can be determined by consulting the manufacturer.

After determining the minimum memory requirements for the appplication, it would be wise to add an additional 25% to 50% more memory. This increase will allow for changes, modifications, or future expansion.

## 4-6   MEMORY ORGANIZATION

Memory organization defines how certain areas of memory for a given controller are used. A diagramatic illustration of memory organization is known as a *memory map* (Fig. 4-14). The memory map shows not only what is stored in memory, but also where things are stored according to address. This feature makes the memory map a useful tool when creating the control program.

It is unlikely that two different controllers will have identical memory maps, but a generalization of memory organization is still valid in light of the fact that all programmable controllers have similar storage requirements. In general, all programmable controllers must have memory allocated for the following four items.

- Executive Programs
- Processor Work Area or Scratch Pad
- Data Table
- User Program Memory

The Executive and the Scratch Pad are transparent to the user and, thus, can be considered a single area of memory that for our purpose is labeled *System Memory*. On the other hand, the Data Table, and the User Program areas are both accessible to and required by the user. These areas can be grouped into what is called *Application Memory*. The total memory then consists of two main parts: the System Memory and the Application Memory.

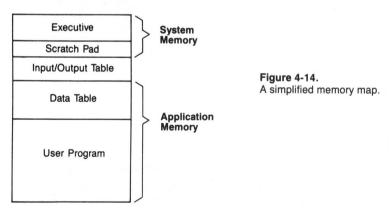

**Figure 4-14.**
A simplified memory map.

It is important to note here that the total memory specified for a controller may include System Memory and Application Memory. For example, a controller with a maximum of 64K may have Executive routines that use 32K and a System work area of 1/4K. This arrangement would leave a total of 31 3/4 K for Application

Memory. Although it is not always the case, the maximum memory, normally specified for a given controller, will only include the Application Memory. Under this assumption, let's take a closer look at the Application Memory.

## The Application Memory

The Application Memory stores programmed instructions and any data that will be utilized by the processor to perform its control functions. A mapping of the typical elements of this area are shown in Fig. 4-15. Each controller has a maximum amount of application memory, which varies depending on the size of the controller. All data is stored in what is called the Data Table, while programmed instructions are stored in the area allocated for the User Program.

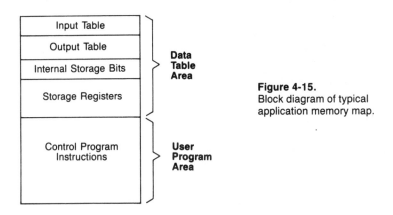

**Figure 4-15.**
Block diagram of typical application memory map.

**Data Table.** The Data Table contents fall primarily into two groups:

1. Status: ON/OFF type information represented by 1s and 0s, stored in unique bit locations.

2. Numbers or codes: Information represented by groups of bits and stored in unique byte or word locations.

The Data Table can be functionally divided into the following areas:

        a) Input Table
        b) Output Table
        c) Internal Storage Bits
        d) Storage Registers

*a) Input Table.* The Input Table is an array of bits that stores the status of digital inputs, which are connected to input interface circuits. The number of bits in the table is equal to the maximum number of inputs. A controller with a maximum of 64 inputs would require an input table of 64 bits. Each connected input has a bit in the input table that corresponds exactly to the terminal to which the input is connected. If the input is ON, its corresponding bit in the table is ON (1). If the input is OFF, the corresponding bit is cleared or turned OFF (0).

    The Input Table is constantly being changed to reflect the current status of the connected input devices. This status information is also being used by the control program. The Input Table is illustrated in Fig. 4-16.

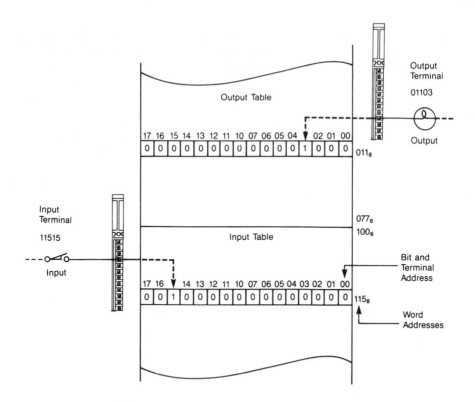

**Figure 4-16.** Block diagram of typical input/output table.

*b) Output Table.* The Output Table is an array of bits that controls the status of digital output devices, which are connected to output interface circuits. The number of bits in the Output Table is equal to the maximum number of outputs. A controller with a maximum of 128 outputs would require an output table of 128 bits.

Each connected output has a bit in the Output Table that corresponds exactly to the terminal to which the output is connected. Note that the bits in the Output Table are controlled by the processor as it interprets the control program and are updated accordingly, during the I/O scan. If a bit in the table is turned ON (1), then the connected output is switched ON. If a bit is cleared or turned OFF (0), the output is switched OFF. The Output Table is shown in Fig. 4-16.

*c) Internal Storage Bits.* Most controllers also allocate an area for what is known as internal storage bits. These storage bits are also called internal outputs, internal coils, internal control relays, or just internals. The internal output operates just as any output that is controlled by programmed logic; however, the output is used strictly for internal purposes. In other words, the internal output does not directly control an output device. Internal outputs are used for interlocking purposes in the control program, just as control relays are used in hardwired logic.

Internal outputs include timers, counters, and internal control relay instructions of various types (see Chapter 7). Each internal output, referenced by an address in the control program, has a storage bit of that same address. When the control logic is True, the internal (output) storage bit turns ON.

*d) Storage Registers.* The information stored in the Input/Output Table is ON/OFF status information, which is easily represented by 1 or 0. To denote any quantity having a value that cannot be represented by a single 1 or 0, groups of bits or memory words must be used. Memory words that store value type information are also called *storage registers.* In general, there are three types of storage registers: *input registers, holding registers, and output registers.* The total number of registers varies depending on the controller memory size and according to how the Data Table is configured (ratio of data storage area to program storage area)*. Values stored in the storage registers are in a binary, or BCD format. Each register can generally be loaded, altered, or displayed by using the programming unit or a special data entry device, offered by most manufacturers.

Input registers are used to store numerical data received via input interfaces, from devices such as thumbwheel switches, shaft encoders, and other devices that provide BCD input. Analog signals also provide numerical data that must be stored in input registers. The current or voltage signal generated by various analog transmitters is converted by the analog interface. From these analog values, binary representations are obtained and stored in the designated input register. The value contained in the input register is determined by the input device and, therefore, is not alterable from within the contoller or via any other form of data entry.

Holding registers are those required to store variable values that are program generated by instructions (e.g. math, timer, or counter) or constant values that are entered via the programming unit or some other data entry method. Table 4-1 shows typical variable and constants stored in holding registers.

**Table 4-1.** Constants and Variables

| CONSTANTS | VARIABLES |
|---|---|
| Timer Preset Value | Timer Accumulated Value |
| Counter Preset Value | Counter Accumulated Value |
| Loop Control Setpoints | Resultant from Math Operations |
| Other Compare Setpoints | Analog Input Values |
| Decimal Tables (recipes) | Analog Output Values |
| ASCII Characters | BCD Inputs |
| ASCII Messages | BCD Outputs |
| Other Numerical Tables | |

Output registers are used to provide storage for numerical or analog values that control various output devices. Typical devices that receive data from output registers are alphanumeric LED displays, recorder charts, analog meters, speed controllers, and control valves. Output registers are essentially holding registers that are designated "output", because of their particular nature (i.e. controlling outputs).

In addition to the I/O tables and storage register areas, some controllers allocate a portion of the Data Table for storing decimal, ASCII, and binary data. Typically, this table area is used for recipe data, report generation messages, or other data that can be stored or retrieved during program execution.

**User Program Memory.** The User Program memory is an area reserved in the Application Memory for the storage of the control logic. All the PC instructions that control the machine or process are stored here. The addresses of inputs and outputs, whether real or internal, are specified in this section of memory.

When the controller is in the run mode, and the program is executed, the processor interprets these memory locations and controls the bits in the data table

---

* In some controllers, the boundary between the Data Table and User Program is selectable. If the Data Table is increased, more space is available for storage registers.

which correspond to a real or internal I/O. The interpretation of the User program is accomplished by the processor's execution of the Executive Program.

The maximum amount of User program memory available is normally a function of the controller size (i.e. I/O capacity). In medium and large controllers, the User program area is normally flexible by altering the size of the Data Table so that it meets the minimum data storage requirements. In small controllers, however, the User program area is normally fixed.

## 4-7 SUMMARY ON APPLICATION MEMORY

This chapter has presented an analysis of programmable controller memory characteristics regarding memory type, storage capacity, organization, and structure. Particular emphasis has been placed on the Application Memory, which stores the control program and data. Careful consideration must be given to the type of memory, since certain applications require frequent change, while others require permanent storage once the program is debugged. A RAM with battery support may be adequate in most cases, but in others a RAM and optional nonvolatile type memory may be required.

It is important to remember that the memory capacity for a particular controller may not be totally available for application programming. The specified memory capacity may include memory utilized by the Executive routines or the Scratch Pad, as well as the User Program area.

The Application Memory varies in size, depending on the size of the controller. The total area available for the control program also varies according to the size of the Data Table. In small controllers, the Data Table is usually fixed, which means that User Program area will be fixed. In the larger controllers, however, the Data Table size is usually selectable, according to the data storage requirements of the application. This allows the program area to be adjusted to meet the application requirements.

When selecting a controller, care must be given to any limitations that may be placed on the use of the available Application Memory. One controller, for instance, may have a maximum of 256 internal outputs with no restrictions on the number used for timers, counters, or various types of internal outputs. Another controller, however, may have 256 internal outputs available with restrictions to 50 timers, 50 counters, and 156 of any combination of various types of internal outputs. A similar type of restriction could also be placed on data storage registers.

One way to be sure to satisfy memory requirements is first to understand the application requirements for program and data storage and the flexibility required for program changes or on-line data entry. This understanding will allow the user to make a decision on the memory type. Creating the program on paper first will help when evaluating capacity requirements. With the use of a memory map, users should learn what is available for the application and then how the application memory is configured for their use. It is also good to know ahead of time if the application memory is expandable.

## 4-8 THE SYSTEM POWER SUPPLY

The System Power Supply plays a major role in the total system operation. It can very well be considered as the first line manager of system reliability and integrity, for its responsibility is not only to provide internal DC voltage to the system components (i.e. processor, memory, and input/output), but also to monitor and regulate the supplied voltages and warn the CPU if all is not well. The power supply then has the function of supplying well-regulated power and protection for other system components.

## The Input Voltage

Usually, PC power supplies require input from an AC power source; however, a few PCs will accept input from a DC source. Those that will accept a DC source are quite appealing for applications such as off-shore drilling operations, where DC sources are commonly used. The most common requirement, however, is for 120 VAC or 220 VAC, while a few controllers will accept 24 VDC.

Since it is quite normal for industrial facilities to experience fluctuations in line voltage and frequency, an important specification for the PC power supply is to tolerate a 10% to 15% variation in line conditions. For example, a power supply with a line voltage tolerance of + 10% when connected to a 120 VAC source will continue to function properly, as long as the voltage remains between 97 VAC and 132 VAC. A 220 VAC power supply with + 10% line tolerance will function properly, as long as the voltage remains between 194 VAC and 250 VAC. When the line voltage exceeds the upper or lower limits for a specified duration (usually 1-3 AC cycles), most power supplies are designed to issue a shutdown command to the processor. Line voltage variations in some plants could eventually become disruptive and may result in frequent loss of production. Normally, in such a case, a constant voltage transformer can be installed to stabilize line conditions.

**Constant Voltage Transformers.** Good power supplies are designed to tolerate what can be considered normal fluctuations in line conditions, but even the best designed power supply cannot compensate for the especially unstable condition of the line voltage found in some industrial environments. Conditions that cause the line voltage to drop below proper levels vary depending on applications, and even plant locations. Some of the causes are:

- Startup/Shutdown of nearby heavy equipment, such as large motors, pumps, welders, compressors, and air conditioning.
- Natural line losses that vary with distance from utility substations.
- Intra-plant line losses caused by poorly made connections.
- Brownout situation in which line voltage is intentionally reduced by the utility company.

A constant voltage transformer compensates for voltage changes at its input (primary) to maintain a steady voltage at its output (secondary). When operated at less than the rated load, the transformer can be expected to maintain approximately + 1% regulation with an input voltage variation of as much as 15%. The percent regulation changes as a function of the operated load (PC power supply and input devices). The constant voltage transformer then must be properly rated to provide ample power to the load. The rating of the constant voltage transformer, in units of volt-amperes (VA), should be selected based on the worst-case power requirement of the load. A recommended rating for the constant voltage transformer should be obtained from the PC manufacturer. Figure 4-17 illustrates a simplified connection of a constant voltage transformer when applied to programmable controllers.

- The Sola* CVS, "standard sinusoidal," or equivalent constant voltage transformer is suitable for PC applications. This type of transformer incorporates line filters to remove high harmonic content and provide a clean sinusoidal output. Constant voltage transformers that do not filter high harmonics are not

---

* Sola Basic Industries, Elk Grove Village, IL

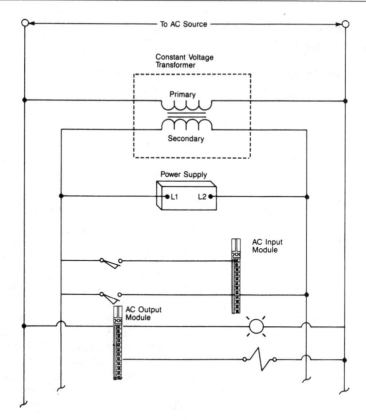

**Figure 4-17.** Simplified connection diagram for a constant voltage transformer.

recommended for PC applications. Figure 4-18 illustrates the relationship between the output voltage and input voltage for a typical Sola CVS transformer operated at different loads.

**Isolation Transformers.** Often, a programmable controller will be installed in an area where the AC line is not unstable, however, surrounding equipment may generate considerable amounts of electromagnetic interference (EMI). Such an installation could result in intermittent misoperation of the controller, especially if the controller is not electrically isolated (on a separate AC power source) from the equipment generating the EMI. Placing the controller on a separate isolation transformer from the potential EMI generators will increase system reliability. The isolation transformer need not be a constant voltage transformer; it should be connected between the controller and the AC power source.

## Loading Considerations

The system power supply provides the DC power for the logic circuits of the CPU and the I/O circuits. Each power supply has a maximum amount of current that it can provide at a given voltage level (e.g. 10 amps at 5 volts). The amount of current that a given power supply is capable of providing is not always sufficient to supply a particular mix of I/O modules. Such a case would cause undercurrent conditions, which could result in unpredictable operation of the I/O system.

The undercurrent situation under most circumstances is unusual, since most power supplies are designed to accommodate a general mix of the most commonly used I/O modules. The undercurrent condition normally arises in large PCs as opposed to small ones, and then when an excessive number of special purpose

I/O modules are used (e.g. power contact outputs, analog inputs/outputs). These special purpose modules usually have higher current requirements than most commonly used digital I/O modules.

Power supply overloading can be an especially annoying condition, since the problem is not always easily detected. The overload condition is often a function of the combination of outputs that are ON at a given time, which means that the overload conditions can appear intermittently. When power supply loading limits have been exceeded, the normal remedy is to add another supply. To be aware of system loading requirements ahead of time, users can obtain from the vendor specifications on typical current requirements of the various I/O modules. Be sure that information provided includes per point (single input or output) requirements and current requirements for ON and OFF states. If the summation of the current requirements for a particular I/O configuration is greater than the total current supplied by the power supply, then a second power supply will be required. An early consideration of line conditions and power requirements will help to avoid problems during installation and start-up.

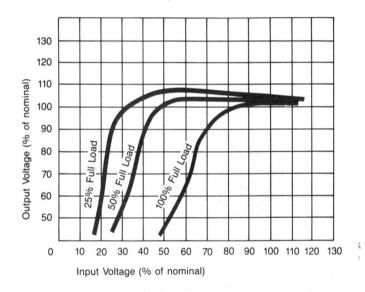

**Figure 4-18.** Illustration of input voltage regulation for a typical Sola unit.

# CHAPTER
# --]5[--

# THE
# INPUT/OUTPUT
# SYSTEM

*The exceptional flexibility and maintainability of the programmable controller is substantially due to the input/output system.*

## 5-1 INTRODUCTION

The input/output (I/O) system provides the physical connection between the outside world (field equipment) and the central processing unit (Fig. 5-1). Through various interface circuits, the controller can sense and measure physical quantities regarding a machine or process, such as proximity, position, motion, level, temperature, pressure, current, and voltage. Based on status sensed or values measured, the CPU issues commands that control various devices such as valves, motors, pumps, and alarms. In short, the input/output interfaces are the sensory and motor skills required by the CPU to exercise control over a machine or process.

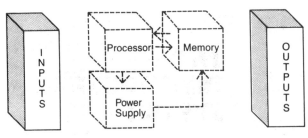

**Figure 5-1.** I/O system.

The early predecessors of today's PCs were limited to discrete input/output interfaces, which only allowed ON/OFF type devices to be connected. This limitation allowed the PC only partial control of many process applications. These applications required analog measurement and manipulation of numerical values to control analog and instrumentation devices. Today's controllers, however, have a complete range and variety of discrete and analog interfaces that allow them to be applied to practically any type of control. The components of a typical I/O system are shown in Fig. 5-2. This chapter will introduce these interfaces, explain their physical, electrical, and functional characteristics, and explore their methods of providing interface to the outside world.

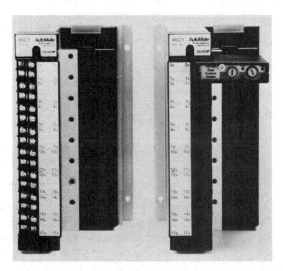

**Figure 5-2.** I/O rail (left), and rail with installed I/O (Courtesy of Reliance Electric Co.).

## 5-2 DISCRETE INPUTS/OUTPUTS

The most common class of input/output interface is the discrete type. This interface connects field input devices, which will provide an input signal that is separate and distinct in nature, or field output devices that will require a separate and distinct signal to control their state. This characteristic limits the discrete I/O interfaces to sensing signals that are on/off, open/closed, or equivalent to a switch closure. To the interface circuit then, all discrete inputs are essentially a switch that is open or closed. Likewise, output control is limited to devices that only require switching to one of two states, such as on/off, open/closed, or extended/retracted. Table 5-1 lists several discrete input/output devices.

**Table 5-1.** Discrete Inputs/Outputs

| Input Devices | Output Devices |
|---|---|
| Selector Switches | Alarms |
| Pushbuttons | Control Relays |
| Photoelectric Eyes | Fans |
| Limit Switches | Lights |
| Circuit Breakers | Horns |
| Proximity Switches | Valves |
| Level Switches | Motor Starters |
| Motor Starter Contacts | Solenoids |
| Relay Contacts | |

### Standard Ratings

Each discrete input and output is powered by some field supplied voltage source that may or may not be of the same magnitude or type (e.g. 120 VAC, 24 VDC). For this reason, I/O interface circuits are available at various AC and DC voltage ratings, as listed in Table 5-2.

**Table 5-2.** Standard Ratings for Discrete I/O Interfaces

| Input Interfaces | Output Interfaces |
|---|---|
| 24 Volts AC/DC | 12-48 Volts AC |
| 48 Volts AC/DC | 120 Volts AC |
| 120 Volts AC/DC | 230 Volts AC |
| 230 Volts AC/DC | 12-48 Volts DC |
| TTL level | 120 Volts DC |
| Non-Voltage | 230 Volts DC |
| Isolated Input | Contact (relay) |
| | Isolated Output |
| | TTL level |

When in operation, if an input switch is closed, the input interface senses the supplied voltage and converts it to a logic-level signal acceptable to the CPU to indicate the status of that device. A logic 1 indicates ON or CLOSED, and a logic 0 indicates OFF or OPEN. In operation, the output interface circuit switches the supplied control voltage that will energize or de-energize the device. If an output is turned ON through the control program, the supplied control voltage is switched by the interface circuit to activate the referenced (addressed) output device.

To apply these input/output interfaces properly, it is important to have a general understanding of how they operate and to have an awareness of certain operating specifications. These specifications are discussed in section 6 of this chapter. Now, let us look at the various discrete interfaces.

## Interface Descriptions

**AC/DC Inputs.** A block diagram of a typical AC/DC input interface circuit is shown in Fig. 5-3. Input circuits vary widely among PC manufacturers, but in general all AC/DC interfaces operate in a manner similar to that described in this diagram. The input circuit is composed of two primary parts: the *power section* and the *logic section*. The power and logic sections of the circuit are normally (but not always) coupled with a circuit, which electrically separates the two.

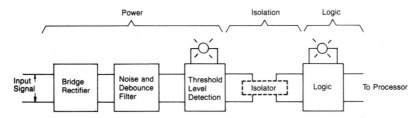

**Figure 5-3.** Block diagram for AC/DC input circuit.

The power section of an input interface basically performs the function of converting the incoming voltage (230 VAC, 115 VAC, etc.) from an input sensing device, such as those described in Table 5-1, to a DC logic-level signal to be used by the processor during its scan. The bridge rectifier circuit converts the incoming signal (AC or DC) to a DC level that is passed through a filter circuit, which will protect against signal bouncing (de-bounce) and electrical noise on the input power line. This filter causes a signal (input) delay that is typically 9-25 msec. The threshold circuit detects whether the incoming signal has reached the proper voltage level for the specified input rating. If the input signal exceeds and remains above the threshold voltage for a duration of at least the filter delay, the signal will be recognized as a valid input.

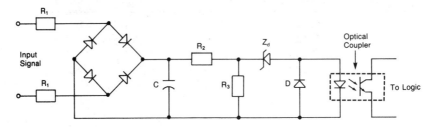

**Figure 5-4.** Typical AC/DC input circuit.

When a valid signal has been detected, it is passed through the *isolation* circuit, which completes the electrically isolated transition from AC to logic level. The DC signal from the isolator is used by the logic circuit and made available to the processor for its data bus. Electrical isolation is provided so that there is *no electrical connection* between the field device (power) and the controller (logic). This electrical separation will help prevent large voltage spikes from damaging the logic side of the interface (or the controller). The coupling between the power and logic sections is normally provided by an optical-coupler or a pulse transformer.

Figure 5-4 illustrates a sample input circuit. Most input circuits will have an LED (power) indicator to signify that the proper input voltage level is present (a switch is closed). An LED indicator may also be available to indicate the presence of logic 1, if the input voltage is present, and if the logic circuit is functioning properly. A device connection diagram is shown in Fig. 5-5.

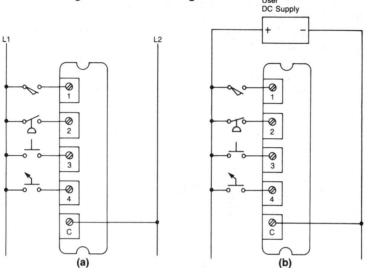

**Figure 5-5.** Typical connection diagram for AC **(a)**, and DC **(b)**.

**TTL Inputs.** The TTL input interfaces allow the controller to accept signals from TTL-compatible devices, including solid-state controls and sensing instruments. TTL inputs are also used for interfacing with some 5 VDC-level control devices and several types of photoelectric sensors. The TTL interface has a configuration similar to the AC/DC inputs; however, the input delay time caused by filtering is generally much shorter. TTL input modules normally require an external $\pm 5$ VDC power supply with certain current specifications. Figure 5-6 shows a typical TTL input connection diagram.

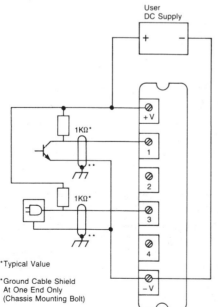

**Figure 5-6.** Typical TTL connection diagram.

**Non-Voltage Input.** The non-voltage input interface accepts discrete inputs, as those mentioned in Table 5-1, but does not require that the field devices be externally powered from a field source. Non-voltage input modules contain an internal power supply (normally 12-24 VDC), which provides the necessary power for sensing the closure of the "dry contact" inputs. The internal interface power source receives power from the system power supply.

**AC Outputs.** Figure 5-7 shows a block diagram of a typical AC output circuit. AC output circuits, like input circuits, vary widely among PC manufacturers. The block configuration, however, describes basic operation of AC outputs. The circuit consists primarily of the logic and power sections, coupled by an isolation circuit. The output interface can be thought of as a simple switch through which power can be provided to control the output device.

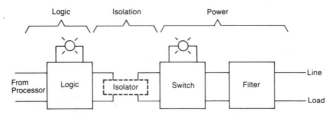

**Figure 5-7.** Block diagram for AC output circuit.

During normal operation, the processor sends to the logic circuit the output status according to the logic program. If the output is energized, the signal from the processor is fed to the logic section and passed through the isolation circuit, which will switch the power to the field device. Figure 5-8 shows a sample AC output circuit.

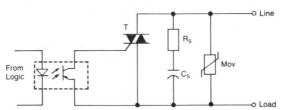

**Figure 5-8.** Typical AC output circuit.

The switching section generally uses a *Triac* or a *Silicon Controlled Rectifier (SCR)* to switch the power. The AC switch is normally protected by an *RC snubber* and often a *Metal Oxide Varistor (MOV)*, which is used to limit the peak voltage to some value below the maximum rating and also to prevent electrical noise from affecting the circuit operation. A fuse may be provided in the output circuit to prevent excessive current from damaging the AC switch. If the fuse is not provided in the circuit, it should be user supplied and should conform to the manufacturer's specification.

As with input circuits, the output interface may provide LED indicators to indicate operating logic and power circuits. If the circuit contains a fuse, a fuse status indicator may also be incorporated. An AC output connection diagram is illustrated in Fig. 5-9. Note that the switching voltage is field supplied to the module. Several factors that must be considered when connecting AC outputs are discussed in Chapter 10.

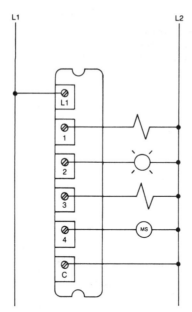

**Figure 5-9.** Typical AC output connection diagram with common return.

**DC Outputs.** The DC output interface is used to switch DC loads. Functional operation of the DC output is similar to the AC output; however, the power circuit generally employs a *power transistor;* to switch the load. Like triacs, transistors are also susceptible to excessive applied voltages and large surge currents, which could result in overdissipation and a short circuit condition. To prevent this condition from occurring, the power transistor will normally be protected by a *free wheeling* diode. Figure 5-10 shows a typical DC output circuit and device connection.

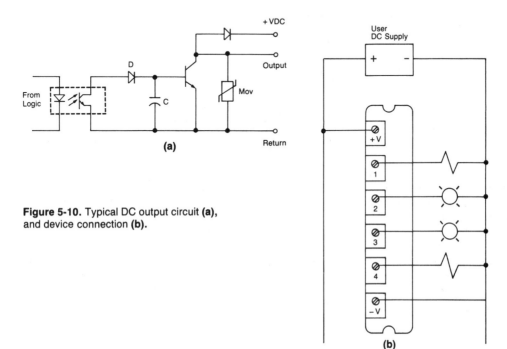

**Figure 5-10.** Typical DC output circuit **(a)**, and device connection **(b)**.

**Contact Outputs.** The contact output interface allows output devices to be switched by a N.O. or N.C. relay contact. Electrical isolation between the power output signal and the logic signal is provided by the separation not only between the contacts, but also between the coil and contacts. Filtering, suppression, and fuses are also generally incorporated.

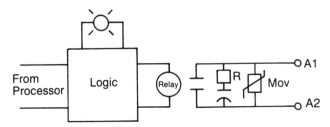

**Figure 5-11.** Typical contact output circuit.

The contact output can be used to switch either AC or DC loads, but are normally used in applications such as multiplexing analog signals, switching small currents at low voltage, and interfacing with DC drives for controlling different voltage levels. High power contact outputs are also available for applications which require switching of high currents. A contact output circuit is shown in Figure 5-11. The device connection for this output module is similar to the AC output.

**TTL Outputs.** The TTL output interface allows the controller to drive output devices that are TTL compatible, such as seven-segment LED displays, integrated circuits, and various 5 VDC devices. These modules generally require an external ±5 VDC power supply with specific current requirements. Figure 5-12 illustrates typical output connections to TTL compatible devices.

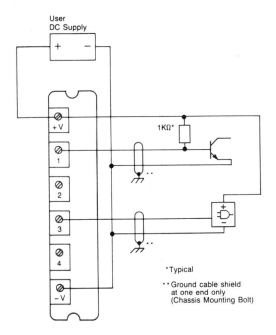

**Figure 5-12.** Typical TTL output circuit.

82

**Isolated Inputs/Outputs.** Input and output interfaces usually have a common return line connection for each group of inputs or outputs on a single module. However, sometimes it may be required to connect an input or output device of different ground levels to the controller. In this case, isolated input or output interfaces with *separate return lines* for each input/output circuit are available (AC or DC) to accept these signals. The operation of the isolated interface is the same as the standard discrete I/O. Figure 5-13 illustrates sample device connections for an isolated interface.

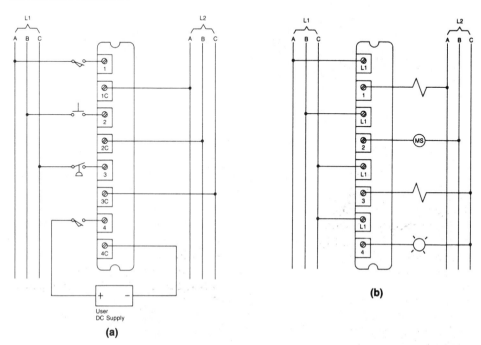

**Figure 5-13.** I/O connection diagram for **(a)** isolated inputs for AC and DC, **(b)** isolated outputs.

## 5-3 NUMERICAL DATA INPUT/OUTPUT INTERFACES

With the integration of the microprocessor into PC architecture in the early 1970s came new capabilities for arithmetic operation and data manipulation. This expanded processing capability led to a new class of input/output interfaces known as numerical data I/O. Numerical input interfaces allowed measured quantities to be input from instruments and other devices that provide numerical data, while numerical output interfaces allowed control of devices that require numerical data.

In general, numerical data I/O interfaces can be categorized into two groups: those that provide interface to *multi-bit* digital devices and those that provide interface to *analog* devices. The multi-bit interfaces are like the discrete I/O in that the processed signals are discrete. The difference, however, is that with the discrete I/O only a single bit is required to read an input or control an output. Multi-bit interfaces allow a group of bits to be input or output as a unit to accommodate devices that require the bits be handled in parallel form (BCD inputs or outputs) or in serial form (pulse inputs or outputs). The analog I/O will allow monitoring and control of analog voltages and currents, which are compatible with many sensors, motor drives, and process instruments. With the use of multi-bit or analog I/O, most process variables can be measured or controlled with appropriate interfacing. Table 5-3 lists typical devices that are interfaced to the controller with analog or multi-bit interfaces.

**Table 5-3.** Typical Numerical Input/Output Devices

| Analog Inputs | Analog Outputs |
|---|---|
| Temperature Transducer<br>Pressure Transducers<br>Load Cell Transducers<br>Humidity Transducers<br>Flow Transducers<br>Potentiometers | Analog Valves and Actuators<br>Chart Recorders<br>Electric Motor Drives<br>Analog Meters |
| Multi-Bit Inputs | Multi-Bit Outputs |
| Thumbwheel Switches<br>Bar Code Readers<br>Encoders | Seven-Segment Displays<br>Intelligent Displays |

## Standard Ratings

Analog I/O interfaces are generally available for several standard *unipolar* (positive swing only) and *bipolar* (negative and positive swings) ratings. In most cases, a single input or output interface can accommodate two or more different ratings and can satisfy either a current or voltage requirement. The different ratings will be either hardware (rocker switches or jumpers) or software selectable. Table 5-4 lists the standard analog I/O ratings.

**Table 5-4.** Standard Analog I/O Ratings

| Input Interfaces | Output Interfaces |
|---|---|
| 4-20 mA | 4-20 mA |
| 0 to +1 Volts DC | 10-50 mA |
| 0 to +5 Volts DC | 0 to +5 Volts DC |
| 0 to +10 Volts DC | 0 to +10 Volts DC |
| 1 to +5 Volts DC | ± 2.5 Volts DC |
| ± 5 Volts DC | ± 5 Volts DC |
| ± 10 Volts DC | ± 10 Volts DC |

## Interface Descriptions

**Analog Inputs.** The analog input interface contains the circuitry necessary to accept analog voltage or current signals from field devices. The voltage or current inputs are converted from an analog to a digital value by an *Analog-to-Digital Converter (ADC)*. The conversion value, which is proportional to the analog signal, is passed through to the controller's data bus and stored in a memory location for later use.

Typically, analog input interfaces have a very high input impedance, which allows them to interface to high source-resistant outputs from input devices. The input line from the analog device generally uses shielded conductors. The shielded cable provides better interface media and helps to keep line impedance imbalances low, in order to maintain good common mode rejection of noise levels, such as power line frequencies. The input stage of the interface provides filtering and isolation circuits to protect the module from additional field noise.

The analog value (after conversion) is expressed as a BCD value that ranges from 0000 to 9999 or a decimal value from 0 to 32767, where the low and high counts represent the low and full scale input signals. A typical analog input connection is illustrated in Fig. 5-14.

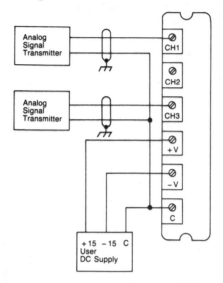

**Figure 5-14.** Typical analog input connection diagram.

**Analog Outputs.** The analog output interface receives from the processor numerical data, which is translated into a proportional voltage or current to control an analog field device. The digital data is passed through a *Digital-to-Analog Converter (DAC)* and output in analog form. Isolation between the output circuit and the logic circuit is generally provided through optical-couplers. These output interfaces normally require an external power supply with certain current and voltage requirements. Figure 5-15 illustrates a typical device connection for an analog output interface.

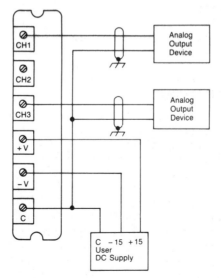

**Figure 5-15.** Typical analog output connection diagram.

**Register (BCD) Inputs.** The register or BCD input interface provides *parallel communication* between the *processor* and *input devices,* such as thumbwheel switches. This interface is generally used to input parameters into specific register locations in memory to be used by the control program. Typical parameters are timer and counter presets, and set point values.

These interfaces generally accept voltages in the range of 5 VDC (TTL) to 24 VDC and are grouped in a module containing 16 or 32 inputs, which correspond to one or two I/O registers. Data manipulation instructions, such as GET or Block Transfer In, are used to access the data from the register input interface. Figure 5-16 illustrates a typical device connection for a register input.

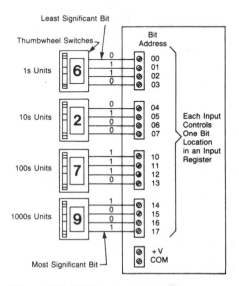
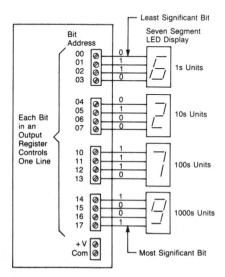

**Figure 5-16.** BCD input connection diagram for register input module.

**Figure 5-17.** BCD output connection diagram for register output module.

**Register (BCD) Outputs.** This numerical interface provides *parallel communication* between the *processor* and an *output device,* such as a seven segment LED display or an BCD alphanumeric display. The register (BCD) output interface can also be used to drive small DC loads that have low current requirements (0.5 amp). The register output interface generally provides voltages that range from 5 VDC (TTL) to 30 VDC and have 16 or 32 output lines (one or two I/O registers).

When information is sent from the processor through a data transfer or I/O register instruction, the data is latched in the module and made available at the output circuit. Figure 5-17 shows a typical register output device connection.

**Encoder/Counter Input.** The encoder/counter interface provides a *high speed counter,* external to the processor, which responds to input pulses sensed at the interface. This counter is, in general, independent of either program scan or I/O scan. Typical applications of the encoder/counter interface are operations that require direct encoder input to a counter, which is capable of providing direct comparison outputs. Such operations are in closed-loop positioning of machine tool axes, hoists, and conveyors, as well as in cycle monitoring on high speed machines such as can making equipment, stackers, and forming equipment.

The encoder/counter input module accepts input pulses from an incremental encoder, which provides pulses that signify position when the encoder rotates. The pulses are counted and sent to the processor. Absolute encoders are generally used with interfaces that receive BCD or Gray code data, which represents the angular position of the shaft.

During operation, the module receives pulse inputs, which are counted and compared with a user specified preset value. The counter input module may have an output line available, which is energized when the input and preset counts are equal. The maximum pulse frequency varies and ranges from 100 Hz to 50 KHz with minimum pulse width of 10 to 20 microseconds.

These interfaces can receive pulses in a single channel or dual channel form from a user-supplied encoder. Typically, the module provides a TTL or DC (24V) output and accepts a 48 VDC *limit switch* and *marker* inputs. The outputs normally represent conditions such as *accumulated count* equal to *preset count* or *accumulated count* greater than the *preset count*.

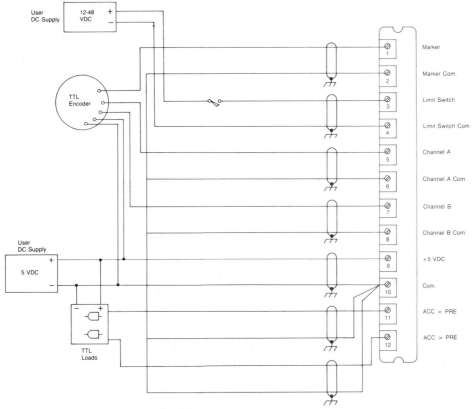

**Figure 5-18.** Typical encoder/counter module connection diagram, with TTL loads connections to accumulated count equal to preset and accumulated count greater than preset.

The communication between the encoder/counter interface and the processor is bi-directional. The module accepts the preset value and other control data from the processor and transmits values and status data to the PC memory. The output controls are enabled from the control program, which tells the module to operate the outputs according to the count values received. The processor, also through the program, enables and resets the counter operation. The interface can let the PC know when the marker and limit switch are both energized indicating a home situation (position).

Typically, the maximum length between the module and the encoder should not exceed 50 feet, and shielded cables are generally used. The interface provides isolation between input/output circuits as well as between the control logic and both I/O circuits. The isolation is enhanced by the use of separate power supplies that must be provided by the user. (See Fig. 5-18).

## 5-4    SPECIAL I/O

In the previous sections, we discussed standard discrete, analog, and numerical I/O interfaces that will probably cover the requirements for at least 90% of the applications that the user will encounter. To process certain types of signals or information efficiently, the controller will require special interfaces. These special I/O interfaces, sometimes called *preprocessing* modules, include those that simply condition input signals, such as low-level analog, fast inputs, or other signals that cannot be interfaced using standard I/O modules. Special I/O modules may also incorporate an on-board microprocessor to add intelligence to the interface. These *intelligent* modules can perform complete processing tasks, independent of the CPU (and program scan). This method of allocating various control tasks to I/O interfaces is known as *distributed processing*. In this section, we will discuss the most commonly available special interfaces. Although these modules are usually supported by the medium to very large PCs, distributed intelligence will eventually be supported by smaller controllers.

### Interface Descriptions

**Thermocouple Input.** In addition to the standard analog voltage or current input interfaces that receive signals from transmitters, special analog inputs can accept signals directly from the transducer. One example is the thermocouple input interface, which accepts type J thermocouple outputs and provides cold junction compensation on board to correct changes in cold junction temperatures. The operation of this interface is similar to the standard analog input with the exception that very low level signals are acceptable from the thermocouple device (approximately 43 mV at maximum temperature). These signals are filtered, amplified, and digitized through the A/D converter and then sent to the processor via the I/O bus on command from a program instruction. The data is used by the control program to execute temperature related control. Figure 5-19 illustrates typical device connections for these analog interfaces.

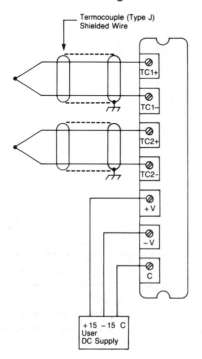

**Figure 5-19.** Typical thermocouple input connection diagram.

**Fast Input.** The fast response input interface is used to detect input pulses of very short duration. Certain devices generate signals that are much faster than the PC scan time and cannot be detected through regular I/O modules. The fast response input interface operates as a *pulse stretcher* to enable the input signal to remain valid for one scan. If the controller has immediate input capabilities, it can be used to respond to a fast input that could initiate an interrupt routine in the control program.

The input voltage range is normally between 10 and 24 VDC, and the logic data signal can be activated by the leading or trailing edge of the triggering input. When the input is triggered, the interface stretches the input signal and has it available for the processor I/O bus. Filtering and isolation are also provided; however, the filtering causes a very short input delay since the normal input devices connected to this interface do not have contact bounce. Typical devices include proximity switches, photoelectric cells, or instrumentation equipment that provides pulse signals with durations ranging between 50 to 100 microseconds.

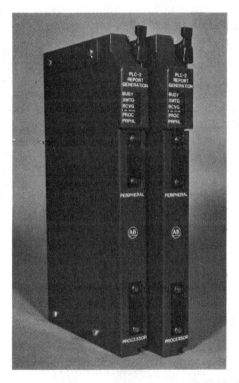

**Figure 5-20.** Report generation (ASCII) module (Courtesy Allen-Bradley Co.).

**ASCII.** The ASCII input/output interface is used for sending and receiving alphanumeric data between peripheral equipment and the controller. Typical peripheral devices include printers, video monitors, and displays. This special I/O interface, depending on the manufacturer, is available with communications circuitry only or complete communication interface circuitry that includes on-board RAM buffer and a dedicated microprocessor. The information exchange of either type interface generally takes place via an RS-232C, RS-422, or a 20 mA current loop standard communications link (see Chapter 6 for peripheral interfacing). An ASCII interface generally requires an isolated power supply. The Allen-Bradley report generation (ASCII) module is shown in Fig. 5-20.

If the ASCII interface does not use a microprocessor, the main processor handles all the communications handshaking, which significantly slows the communication process and the program scan. Each character or string of characters to be transmitted to the module or received from the module is handled on a character-by-character (interrupt) basis. The module interrupts the main CPU each time it receives a character from the peripheral, and the CPU accesses the module each time it needs to send a message to the peripheral. The communication speed is generally very slow, and for a character to be read, the scan time must be faster than the time required to accept one character. For example, if a scan time is 20 msec, and the baud rate is 300 (30 characters per second), a character will be received every 33.3 msec. Conversely, if the baud rate is 1200 (120 characters per second), more than one character per scan will be transmitted from the peripheral (one character every 8.33 msec). If this is the case, it is obvious that several characters will be lost.

If a smart ASCII interface is used, the transmission is accomplished between the peripheral and the module also on an interrupt basis, but at a faster transmission speed. This is possible since the on-board microprocessor is dedicated to performing the I/O communication. The on-board microprocessor contains its own RAM memory, which can store blocks of data that is to be transmitted. When the input data from the peripheral is received at the module, it is transferred to the PC memory through a data transfer instruction at the I/O bus speed. All the initial communication parameters, such as number of stop bits, parity (even or odd) or non-parity, and baud rate, will be hardware selectable (using rocker switches or jumpers) or selectable through control software depending on the interface. This method significantly speeds up the communication process and increases data throughput.

**Figure 5-21.** Strain gauge module (rights) and exitation module (Courtesy of Cincinnati Milacron Co.).

**Strain Gage Amplifier.** This special module (shown in Fig. 5-21), developed by Cincinnati Milacron for its Acramatic Programmable Controller (APC) system, is used to interface with pressure transducers that employ strain gage circuits. The output signal of these transducers has a very low level and needs to be amplified to a usable signal. The strain gage module can receive signals from two strain gages and have two outputs, which have been amplified. The amplified signals can be input to an analog input module as a pressure control variable for monitoring and control. The amplifier module provides separate SPAN and ZERO adjustments for field calibration, if necessary. The strain-gage requires a precision ± 10 VDC level excitation from an excitation module, also provided by Cincinnati Milacron.

**Stepper Motor (Pulse) Output.** The stepper motor interface generates a pulse train that is compatible with stepper motor translators. The pulses sent to the translator will represent distance, rate, and direction commands to the motor.

The stepper motor interface accepts position commands from the control program. Position is determined by the preset count of output pulses, a forward or reverse direction command, and the acceleration or deceleration command for ramping control. The acceleration or deceleration is determined by the rate of output pulses. These commands are generally specified during program control, and once the output interface is initialized by a start, it will output the pulses according to the predefined commands. Once the motion has started, the output module will generally not accept any commands from the processor until the move is completed (preset position = current position). Some interfaces may offer an override command that will reset the current position, which must be disabled to continue operation. The module sends bit information regarding its status to the main processor.

The step rate of pulses can range from 1 to 60 KHz with selectable step pulse widths. These interfaces generally require an external power supply. Figure 5-22 illustrates a typical interface connection to a stepper motor device.

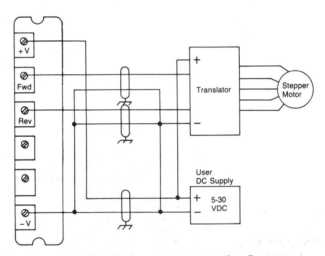

**Figure 5-22.** Typical stepper motor connection diagram.

**Servo Interface.** This interface is being used in applications that once employed clutch-gear systems or other mechanical arrangements to perform motion control. The advantages of servo control are: shorter positioning time, higher accuracy, better reliability, and improved repeatability in the coordination of axes motion. The capabilities of these modules have allowed some programmable controllers to perform functions such as point-to-point control and axis positioning, using servo mechanisms that once required computer numerical control (CNC) machines. Low end CNC users can now specify a PC with servo control, instead of a CNC, at approximately one third the cost.

Programmable controllers that support the servo function module generally require two or more interfaces to implement the servo control task and close the servo loop. The interfaces used combine the feedback pulse from a resolver/encoder to generate an error voltage, which is used through a D/A converter to drive a DC motor. The encoder receives the *Quadrature* (cosine wave) phase which is decoded into forward or reverse count direction and is compared with the drive count (speed).

The Giddings & Lewis servo positioning configuration uses three modules to achieve the proportional closed loop servo control. These modules are the *D/A converter, Digitizing Reference,* and *Microprocessor Digitizing* interfaces. Figure 5-23 illustrates a proportional servo control with resolver feedback for a one-axis system. The D/A converter interface outputs a DC signal to the servo drive based on the endpoint position, axis velocity, and feedback from the resolver. The Digitizing Reference module provides two excitation signals to the resolver; these are the *Reference* (sine wave) and *Quadrature* (cosine wave) which are 90 degrees out of phase from one another. The signal frequency is 2 KHz (Fig. 5-24). The Microprocessor Digitizing module accepts the signals from the resolver, which are phase shifted with respect to the reference signal and dependent on the resolver's rotation. For each complete resolver rotation, the feedback signal shifts 360°. This module utilizes the phase shift to obtain a digital value, which represents the actual position of the motor (Fig. 5-25).

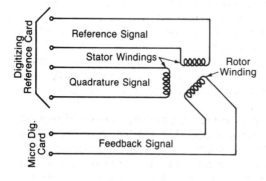

**Figure 5-23.** Proportional servo control for one-axis system (Courtesy of Giddings & Lewis Co.).

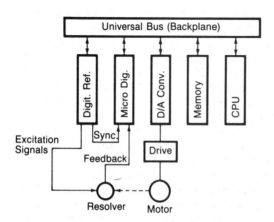

**Figure 5-24.** Exitation signals provided to resolver: Reference and Quadrature (Courtesy of Giddings & Lewis Co.).

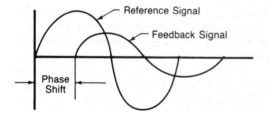

**Figure 5-25.** Phase shifted feedback signal (Courtesy of Giddings & Lewis Co.).

Another manufacturer that provides servo control is Cincinnati Milacron. Their servo control approach uses two modules: the *Feedrate (FDM)* and *DC Motor (DCM)* interfaces. The DCM module combines the command pulses from the FDM module with the feedback pulses from a shaft encoder to generate the error voltage, which is used to interface with a DC servo drive. The FDM module generates the pulses (to the DCM) where the number of pulses represents a distance of movement, the frequency corresponds to a velocity, and the rate of change of frequency represents the acceleration. The feedback data from the encoder is received from the Quadrature phase, which is decoded into forward or reverse count direction in the FDM module. The feedback information is computed in the FDM by an on-board microprocessor, which determines the actual velocity and direction being generated and accelerates towards the preset velocity at the rate programmed in an internal ramp (rate) register.

Figure 5-26 illustrates a typical servo application using the DCM module. The FDM module must be adjacent to the DCM module to pass information through the bus.

**Axis Positioning Module.** The Axis Positioning Module (APM), developed by General Electric for their Series Six programmable controllers, is another special interface that is used for servo position control, in addition to the standard sequencing and control capabilities. The APM interface utilizes an on-board microprocessor, with a 2.34 msec scan time, which controls axis positioning independently of the main CPU. Contrary to other servo control modules, the APM is a single module capable of receiving feedback information from resolvers and controlling the velocity of a servo-motor (through a servo drive) for each axis. This interface receives the set-up data parameters, such as velocity feedforward, position loop gain, and reversal compensation, during initialization at power up.

The APM maintains communication with the main CPU to report the status of machine, module, and total operation of the loop in real time. Alterations of the APM commands can be made under program control if necessary. One outstanding feature of this interface is the 60 error conditions tested by the module and sent back to the CPU. The APM can provide position commands that allow coverage for distances up to 1400 feet at one thousandth (0.001) of an inch resolution and velocity resolution of 0.0001 at 600 inches per minute (10 inches per second). The module can store up to 10 motion profiles, each with up to 50 parameters.

This servo interface is provided with a $\pm$ 10VDC velocity command, Drive OK input, Drive Enable output, Loop Contactor output, up to three resolver excitation outputs and inputs, limit switch inputs for Left End of Travel and Right End of Travel, Home Position, and a Synchronization output for connection with other APMs if required. Figure 5-27 illustrates a typical servo application and a block diagram of the APM I/O connections.

**PID Module.** The *Proportional-Integral-Derivative* interface is used in process applications in which closed-loop control employing the PID algorithm is required. This module provides proportional, integral, and derivative control action from sensed parameters, such as pressure and temperature, which are the input variables to the system. The PID interface is typically applied to any process control operation which requires continuos closed-loop control. PID control is often referred to as a *three-mode close-loop feedback control.*

The basic function of closed-loop process control is to maintain certain process characteristics at desired set points. Generally, the process deviates from the desired set point reference as a result of load material changes and interaction with other processes. During this control, the actual condition of the process characteristics (liquid level, flow rate, temperature, etc.) is measured as a *process variable (PV)* and compared with the target *set point (SP)*. When deviations occur, an error is generated by the difference between the process variable (actual value)

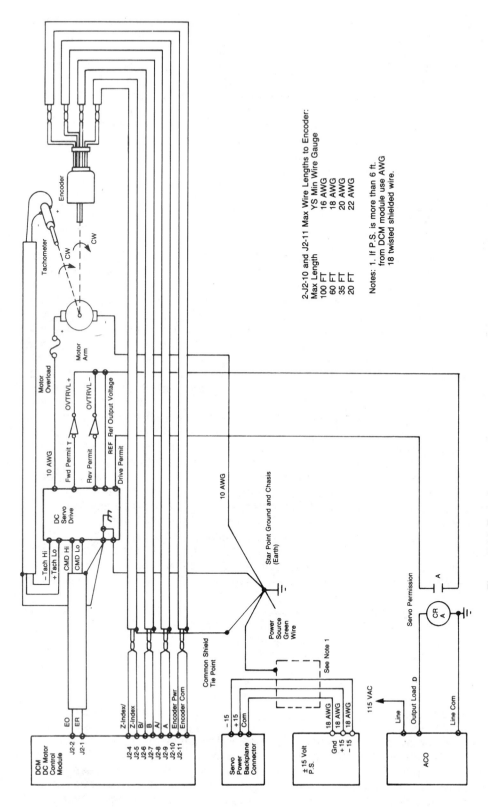

**Figure 5-26.** Typical servo application using the FDM and DCM modules (Courtesy of Cincinnati Milacron Co.).

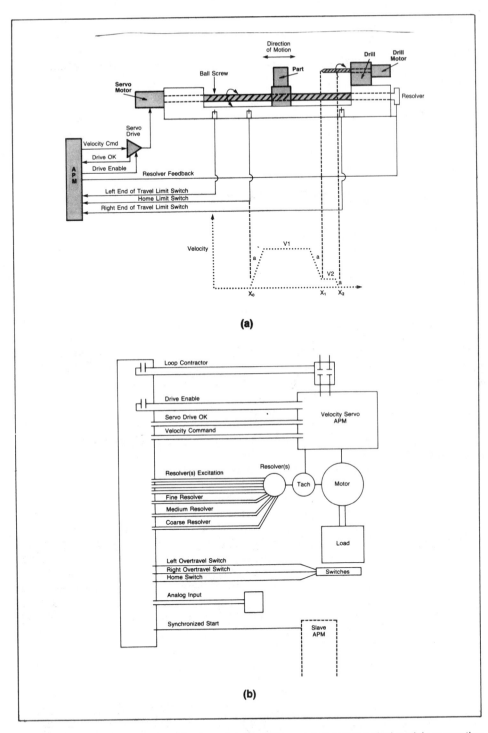

**Figure 5-27.** Axis positioning application using the APM module **(a)**, and typical module connection wiring (Courtesy General Electric Co.).

95

and the set point (desired value). Once an error is detected, the function of the control loop is to modify the *control variable (CV)* output in order to force the error to zero.

The control algorithm implemented in the module is represented by the following equation:

$$Vout = K_P E + K_I \int E \, dt + K_D \frac{dE}{dt}$$

where:

$K_P$ is the proportional gain

$K_I = \dfrac{K_P}{T_i}$ is the integral gain ($T_i$ = reset time)

$K_D = K_P \, T_d$ is the derivative gain ($T_d$ = rate time)

$E = PV—SP$ is the error

Vout is the control variable output

The module receives the process variable in analog form and computes the error difference. The error difference is used in the algorithm computation to provide corrective action at the control variable output. The function of the control action is based on an output control, which is proportional to the instantaneous error value. The *integral control action* (reset action) provides additional compensation to the control output, which causes a change in proportion to the value of the error over a period of time. The *derivative control action* (rate action) adds compensation to the control output, which causes a change in proportion to the rate of change of error. These three modes are used to provide the desired control action in Proportional (P), Proportional-Integral (PI), or Proportional-Integral-Derivative (PID) control fashion.

The information that is sent to the module from the main processor is primarily the control parameters and set points. Depending on the module used, data can be sent to describe the *update time*, which is, the rate or period in which the output variable is updated and the *error deadband*, which is a quantity that is compared to the error signal. If the error deadband is less than or equal to the signal error, no update takes place. Some modules provide square root calculations of the process variable, which is used to perform a *square root* extraction to obtain a linearized scaled output for use by the PID loop. An example application in which the square root extractor can be used is in the PID control of flow. Other parameters, such as maximum error and maximum/minimum control variable output for hi and lo alarms, can also be transmitted to the module if these signals are provided. During operation, the PID interface maintains status communication with the main CPU, exchanging module and process information.

Figure 5-28 illustrates a block example of the PID algorithm for this interface and a typical module connection arrangement.

**Data Processing Modules.** These modules are microprocessor based and perform data processing and file handling functions that would typically be performed by the main CPU or a small dedicated computer. These functions include storing and retrieving recipes, storing and displaying operator messages, generating production reports, and other processing jobs. Allowing these tasks to be done by intelligent modules relieves the main CPU of this burden and helps to uncomplicate the user control logic program by eliminating the typically large amounts of data processing programming.

Two such intelligent data processing units are one offered by Square D Com-

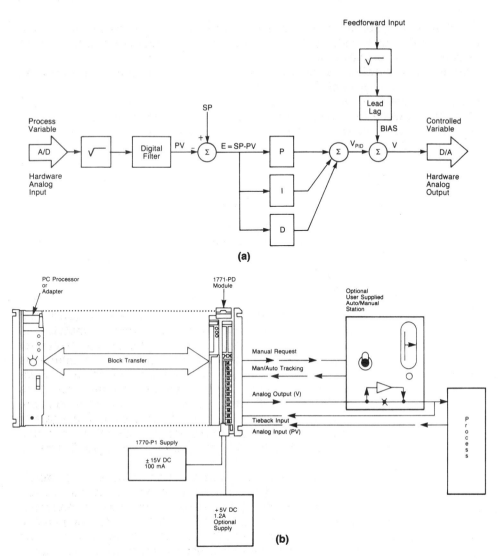

**Figure 5-28.** Block example of PID algorithm **(a)**, and PID module arrangement diagram for A-B model 1771 PID module (Courtesy of Allen-Bradley Co.) **(b)**.

pany for use with the SY/MAX controllers, and another is offered by General Electric Company, for use with the Series Six programmable controllers. The SY/MAX Data Controller Module, as it is called, plugs into any register I/O slot and can communicate with Square D PC processors either directly or over the SY/NET communications network. This module is programmed using the Extended BASIC language and is standard with 9K bytes of user memory, which is expandable up to 300K bytes. The GE Data Processing Unit (DPU) is compatible with their current models 60, 600, and 6000 PCs; however, an additional rack is required to house the DPU. A total of eight communication ports are available—one programming port and seven for general purpose serial communications. The DPU can provide memory storage of up to 256K RAM.

**Network Interface Module.** Network interface modules are designed to allow a number of PCs and other intelligent devices to communicate over a high-speed local area communication network (see Chapter 11). Normally the PCs that can interface to the network are restricted to products designed by the network manufacturer. The other types of devices that can be interfaced and the total number of devices that can be connected are dependent on the network interface.

In general, when a message is sent by a processor or other network device, its network interface retransmits the message over the network at whatever the network baud rate is. The receiving network interface accepts the transmission and sends it to the intended device. The protocol for the communication link varies depending on the network.

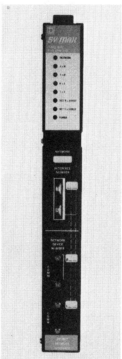

**Figure 5-29.** Network interface module (Courtesy of Square D Co.).

Figure 5-29 shows the SY/NET (Square D Company) network interface module. The module allows up to 200 Square D controllers and other devices to be connected using twin-axial cable, at distances up to 9,000 feet. The interface will allow the CRT programmer to program any PC on the network. The baud rates for the communication ports on each interface module are adjustable, from 300 to 9600 in order to accommodate printers, modems, and data terminals. The network interface module can also be used to connect two or more networks together. Note in Fig. 5-29 that this network interface module has two communication ports to allow two devices to communicate with each other, as well as with the network.

## 5-5   REMOTE I/O

Larger PC systems (normally upward of 512 I/O) will allow input/output subsystems to be remotely located from the central processing unit (but still under its control). A remote subsystem is usually a rack-type enclosure in which the I/O modules are installed. The rack generally includes a power supply to drive the logic circuitry of the interfaces and a remote I/O adapter module which allows communication with the processor. Capacity of a single subsystem is normally 32, 64, 128, or 256 I/O points. A large system with a maximum capacity of 1024 I/O points might have subsystem sizes of either 64 or 128 points, in which case there could be either eight racks with 128 I/O, sixteen racks with 64 I/O, or some combination of both sizes. Although there are a few exceptions, most controllers with remote I/O allow only discrete interface modules to be placed in the rack.

Individual racks are normally connected to the CPU using a *daisy chain* or *star* configuration (Fig. 5-30), via one or two twisted-pair conductors or a single coaxial cable. The distance the remote rack can be placed away from the CPU varies among products, but can be as much as two miles. Another approach, as taken by the Struthers Dunn Company with their Director 4001, is a fiber-optic data link intended to provide greater distances and higher noise immunity.

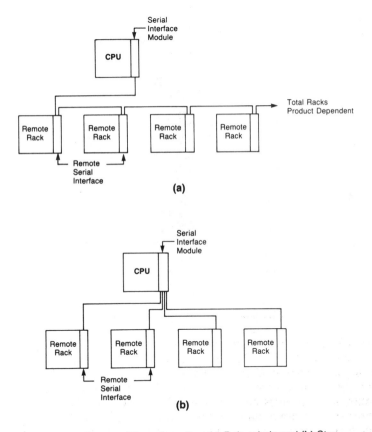

**Figure 5-30.** Remote I/O configuration, **(a)** Daisy chain and **(b)** Star.

Remote I/O offers tremendous savings on wiring materials and labor costs for large systems in which the field devices are in clusters at various spread-out locations. With the CPU in a main control room or some other central area, only the communication link is brought back to the processor, instead of hundreds of field wires. Distributed I/O also offers the advantage of allowing subsystems to be installed and started up independently, as well as allowing maintenance on individual subsystems while others continue to operate.

## 5-6   INTERPRETING THE SPECIFICATIONS

Perhaps with the exception of standard input/output current and voltage ratings, specifications, for I/O circuits are all too often treated as a meaningless listing of numbers. Manufacturers specifications, however, provide much information that defines how a particular interface is correctly and safely applied. The specifications place certain limitations not only on the module, but also on the field equipment that it can operate. Failure to adhere to these specifications could result in misapplication of the hardware, leading to faulty operation or equipment damage. The following specifications should be evaluated for each PC application.

### Electrical

**Input Voltage Rating.** This nominal AC or DC value defines the magnitude and type of signal that will be accepted. Deviation from this nominal value will usually be accepted between plus or minus 10% and 15%. This specification may also be stated as the input voltage range. In this case, the minimum and maximum acceptable working voltages for continuous operation are given. An input circuit rated at 120 VAC may accept a signal of 97 to 132 VAC.

**Input Current Rating.** This nominal value defines the minimum input current (at the rated voltage) that the input device must be capable of driving to operate the input circuit. This specification may also appear indirectly as the minimum power requirement.

**Input Threshold Voltage.** This value specifies the voltage at which the input signal is recognized as being absolutely ON. This specification is also called the ON threshold voltage. Some manufacturers also specify an OFF voltage, defining the voltage level at which the input circuit is absolutely OFF.

**Input Delay.** The input delay will be specified as a minimum or maximum value. It defines the duration for which the input signal must exceed the ON threshold before being recognized as a valid input. This delay is a result of filtering circuitry provided to protect against contact bounce and voltage transients. The input delay is typically 9-25 msec for standard AC/DC inputs and 1-3 msec for TTL or electronic inputs.

**Output Voltage Rating.** This nominal AC or DC value defines the magnitude and type of voltage source that can be controlled. Deviation from this nominal value is typically plus or minus 10% to 15%. For some output interfaces, the output voltage is also the maximum continuous voltage. The output voltage specification may also be stated    as the output voltage range, in which case, both the minimum and maximum operating voltages are given. An output circuit rated at 48 VDC, for example, may have an absolute working range of 42 to 56 VDC.

**Output Current Rating.** This specification is also known as the *ON-state continuous current rating,* a value that defines the maximum current that a single output circuit can safely carry under load. The output current rating is a function of component electrical and heat dissipation characteristics. This rating is generally specified at an ambient temperature (typically 0-60°C). As the ambient increases, the output current is typically derated. To exceed the output current rating or oversize the manufacturer's fuse rating could result in permanent short circuit failure or other damage.

**Output Power Rating.** This maximum value defines the total power that an output module can dissipate with all circuits energized. The output power rating for a single energized output is the result of multiplying the output voltage rating by the output current rating (e.g. 120V x 2A = 240VA). This value (for a given I/O Module) in units of volt-amps (or watts) may or may not be the same, if all outputs on the module are energized simultaneously. The rating for an individual output when all other outputs are energized should be verified with the manufacturer.

**Current Requirements.** The current requirement specification defines the current demand that a particular input/output module's logic circuitry places on the system power supply. Totaling the current requirements of all the installed modules that one supply supports and comparing the value with the maximum current that can be supplied will indicate if the power supply is adequate. The specification will normally provide a typical rating and a maximum rating (all I/O activated). An undercurrent condition, resulting from insufficient power supply current, can result in intermittent misoperation of field input and output interfaces.

**Surge Current (max).** The surge current, also called *inrush current,* defines the maximum current and duration (e.g. 20 amps. @ .1 sec.) for which an output circuit can exceed its maximum ON-state continuous current rating. Heavy surge currents are usually a result of transients on the output load line or on the power supply line and the switching of inductive loads. Output circuits are normally provided with internal protection by a free wheeling diode, Zener diode, or an RC network across the load terminals; if not, the protection should be provided externally.

**Off State Leakage Current.** Typically this maximum value that defines the small leakage current that flows through the triac/transistor in its OFF state. This value is normally in the order of a few microamperes to a few milliamperes and presents little problem in most cases. It could perhaps present problems in the case of switching very low currents or could give false indication when using a sensitive instrument, such as a volt-ohm meter, to check contact continuity.

**Output ON-Delay.** This specification defines the response time for the output to go from OFF to ON once the logic circuitry has received the command to turn ON. The turn-ON response time of the output circuit will affect the total time required to activate an output device. The actual worst-case time, required to turn the output device ON after the control logic goes TRUE, will be two program scans, plus the I/O update, the output ON-delay, and the device ON-response times.

**Output OFF-Delay.** The OFF-delay specification defines the response time for the output to go from ON to OFF once the logic circuitry has received the command to turn OFF. The turn-OFF response time of the output circuit will affect the total time required to deactivate an output device. The actual worst-case time, required to turn the output device OFF after the control logic goes FALSE, will be two program scans, plus the I/O update, the output OFF-delay, and the device OFF-response times.

**Electrical Isolation.** This maximum value (volts) defines the isolation between the input/output circuit and the controller logic circuitry. Although this isolation protects the logic side of the module from excessive input/output voltages or currents, the power circuitry of the module may be damaged.

**Output Voltage/Current Ranges.** In reference to analog outputs, this specification is typically a nominal expression of the voltage/current swing of the digital-to-analog converter. The output will always be a proportional current or voltage within the output range. A given analog output module may have several hardware or software selectable unipolar or bipolar ranges (e.g. 0 to 10V, -10V to +10V, 4 to 20ma).

**Input Voltage/Current Ranges.** In reference to analog inputs, this specification is typically a nominal expression that defines the voltage/current swing of the analog-to-digital converter. The input will always be a proportional current or voltage within the input range. A given analog input module may have several hardware or software selectable unipolar or bipolar ranges (e.g. 0 to 10V, -10V to + 10V, 4 to 20ma).

**Digital Resolution.** This specification defines how closely the converted analog input/output current or voltage signal approximates the actual analog value within the specified voltage or current range. Resolution is a function of the number of bits used by the A/D or D/A converter. An 8-bit converter has a resolution of 1 part in $2^8$ (1 part in 256). If the range is 0 to 10V, the resolution is 10 divided by 256 (39.06mV/bit).

**Output Fuse Rating.** Fuses are often supplied as a part of the output circuit, but only to protect the semiconductor output device (triac or transistor). The particular fuse that is employed by the interface or recommended (if not incorporated) by the manufacturer has been carefully selected based on the fusing current rating of the output switching device. Fuse rating incorporates a fuse opening time along with a current overload rating, which allows opening within a time-frame that will avoid damage to the triac or transistor. The recommended specifications should be followed when replacing fuses or in the case where fuses are not incorporated into the output.

## Mechanical

**Points Per Module.** This specification simply defines the number of input or output circuits that are on a single encasement (module). Typically, a module will have 1,2,4,8, or 16 points per module depending on the manufacturer. The number of points per module has two implications that may or may not be of importance to the user. First, in general, the less dense (number of points) a module is, the greater the space requirements are; second, the higher the density, the lower the likelihood that the I/O count requirements can be closely matched with the hardware. For example, if the module contains 16 points, and the user requires 17, two modules must be purchased. Thus the user will have purchased 15 extra inputs or outputs. Several different I/O structures are shown in Fig. 5-31 at the end of this chapter.

**Wire Size.** A wire size specification is not always given, but may be of importance to the user and should be verified. It simply defines the number of conductors and the largest wire gauge that the I/O termination points will accept (e.g. 2 # 14 AWG).

## Environmental

**Ambient Temperature Rating.** This value defines what the maximum temperature of the air surrounding the input/output system should be for best operating conditions. This specification takes into consideration the heat dissipation characteristics of the circuit components, which are considerably higher than the ambient temperature rating itself. The ambient temperature is rated much less, so that the surrounding temperature does not contribute excess heat to that already generated by internal power dissipation. The ambient temperature rating should never be exceeded.

**Humidity Rating.** The humidity rating for PCs is typically 0-95% non-condensing. Special consideration should given to insure that the humidity is properly controlled in the area where the input/output system is installed. Humidity is a major atmospheric contaminant, which can cause circuit failure if moisture is allowed to condense on printed circuit boards.

Proper observance of the specifications provided by the manufacturer's data sheets will help to insure good, safe operation of the control equipment. In Chapter 10, other considerations for properly installing and maintaining the input/output system are discussed in greater detail.

**Figure 5-31.**

**(a)** Gould-Modicon model 884 with capacity of handling 256 I/O and modularity of 8-points or 16-points per module.

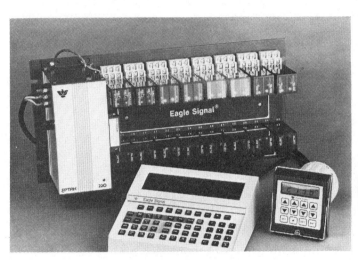

**(b)** Eagle Signal model 220 capable of handling up to 128 I/O on four I/O tracks and modularity of 2 points per module.

**(c)** Allen-Bradley model Mini-PLC 2/15 handles up to 128 I/O with modularity of 8 discrete points per module and 4, 6, or 8 analog points per module.

(d) Square-D model 500 with capacity of 2000 I/O points and modularity of 4 or 8 points per module.

(e) Input/Output rack used with the Westinghouse PC models 700 and 900.

(f) Reliance Electric AutoMate PC model 15 with 64 I/O capacity and modularity of 2 points per module.

# CHAPTER
# --]6[--

# PROGRAMMING AND PERIPHERAL DEVICES

*Many of the major benefits of programmable controllers are realized through the use of peripherals.*

At first glance, most new users of programmable controllers see only their control program stored in memory. As they become more familiar with PCs, it becomes quite clear that their controller's memory contains in various forms a wealth of information, waiting to be extracted. When reproduced in hard copy form, the control program becomes a useful trouble-shooting tool. The status of information regarding each and every connected input and output is also in memory and can be converted to various forms of reports that can aid in maintaining the system. Daily production information can be gathered and output in report form at the end of the day. Converting the information stored in memory into useful information can mean overall enhancement of the control system, while providing helpful and easy-to-use tools for management, maintenance, and operator personnel. In this chapter, we will discuss programming and other peripheral devices that will help the user exploit the full capabilities of the programmable controller.

# 6-1 PROGRAMMING DEVICES

Programmable controller manufacturers have made many advancements in providing the user with better methods of programming that simplify entering, storing, and monitoring the control logic program. Having easy-to-use programming equipment is important, since it represents the primary means by which the user communicates with the PC; simplicity and flexibility play an important role.

The following is a list of the most common peripheral devices associated with PC programming:

- Cathode Ray Tubes (CRTs)
- Mini-Programmers
- Program Loaders
- Memory Burners
- Computers

## Cathode Ray Tubes

CRTs are perhaps the most common devices used for programming the controller. They are self-contained video display units with a keyboard and the necessary electronics to communicate with the CPU and display data. The CRT offers the advantage of displaying large amounts of logic on the screen (Figure 6-1), which greatly simplifies the interpretation of the program. The program logic displayed on the terminal can be ladder diagrams or whatever language is used by the controller.

CRTs are generally classed into two groups: *"dumb"* and *"intelligent"* CRTs. These two types differ greatly in capabilities and in price. Some CRTs are portable and are easily transported through the plant. There are also desk-top types that are used primarily in the office or development laboratory.

**Dumb CRT.** The dumb CRT has been used extensively for years as a relatively inexpensive CRT programming device. As the name implies, this CRT is not microprocessor-based, or "not smart;" all the software needed for creating the program and displaying and updating the screen is contained in the controller's executive memory. This terminal must be connected to the PC to enter or edit the control program. Having to be in constant communication with the processor for programming is known as *on-line* programming. In spite of not being microprocessor based, this dumb video terminal can be used for monitoring and debugging purposes, if the PC executive provides this option.

The dumb CRT offers the advantage of being usable with different makes of PCs. However, all dumb CRTs may not be compatible with a user's particular

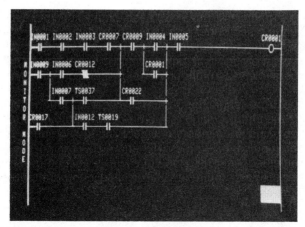

**Figure 6-1.** Relay ladder diagram as shown on a CRT display (Courtesy Westinghouse Electric Co.).

controller. Manufacturers generally provide a list of dumb video terminal models that are recommended for use with their equipment.

**Intelligent CRT.** The intelligent or smart CRT, shown in Figure 6-2, is a microprocessor- based device that not only displays logic networks, but also provides the program editing capabilities and other functions independently of the controller's CPU. Contrasted to the dumb CRT, the intelligent terminal has, internal to its own memory, the software that is required to create, alter, and monitor programs. Intelligent CRTs provide a powerful tool when programming, since all the logic, or control program, can be edited and stored without being connected to the controller. This capability is known as *off-line* programming.

These intelligent devices typically cost two to three times more than the non-smart terminals; however, most manufacturers' designs allow intelligent CRT's to operate with several PCs of their programmable controller family. This intercompatibility among family members makes the purchase of an intelligent CRT more justifiable. In general, these CRTs come with a built-in tape or disk device that permits the permanent storage of one or more programs from different PC sizes. This storage capability can also be used for data collection purposes.

**Figure 6-2.** Intelligent CRT (Courtesy Square D Co.).

More sophisticated smart CRTs have additional features which make them even more appealing. One feature is a network interface that allows the CRT to connect to the manufacturer's Local Area Network (LAN). This arrangement allows the terminal to access any PC in the network, change parameters or programs, and monitor any elements (i.e. coils, contacts, inputs, or outputs) without connecting directly to any specific PC. If the software is available, this arrangement will also allow a means for centralized data collection and display, since it can receive data from the different controllers in the network. Some newer intelligent CRTs provide special graphic keys which allow the user to construct various displays that can be recalled on command. Under program control, variable data can be superimposed on the fixed graphic display. Some intelligent CRTs also have software documentation capabilities, which eliminate the need for purchasing this additional equipment.

As a disadvantage, these smart video terminals are not interchangeable from one manufacturer's PC family to another. However, with the increasing number of products in the manufacturers' product lines and user standardization of products, these intelligent devices will become more economically feasible.

## Mini-Programmers

Mini-programmers, also known as hand-held or manual programmers, are an inexpensive and portable means for programming small PCs. Physically, these devices resemble hand held calculators, but with a larger display and a somewhat different keyboard. The display is usually LED or dot matrix LCD, and the keyboard consists of numeric keys, programming instruction keys, and special function keys. Instead of the hand-held type unit, some controllers have the mini-programmer in a built-in configuration. In some instances, these built-in programmers are detachable from the PC. Figure 6-3 shows a small PC with this built-in and detachable feature. Even though they are used mainly for editing and inputting a control program, mini-programmers can also be a useful tool for starting-up, changing, and monitoring the control logic.

As with CRTs, most manufacturers are now designing mini-programmers that are compatible with two or more controllers of the product family. The mini-programmer is generally used with the smallest member of the family or in some cases with the next larger member, which is normally programmed with a CRT. With this programming option, small changes or monitoring, which may be re-

**Figure 6-3.** Small PC with built-in and detachable mini-programmer (Courtesy of General Electric Co.).

quired in the larger controller, can be accomplished without carrying the CRT to the PC location. This compatibility feature, offered by some mini-programmers, can be translated into cost savings in the case where an expensive CRT cannot be justified.

Mini-programmers, like CRTs, can be non-smart or smart. The non-smart hand-held programmer can be used for entering and editing the program with limited (limited by memory and display size) on-line monitoring and editing capability. The smart mini-programmer is microprocessor-based and provides the user with many of the features offered by the CRT. These smart devices can often perform system diagnostic routines (memory, communication, display, etc.) and even serve as an operator interface device which displays English messages concerning the controlled machine or process. Figure 6-4 shows both types of mini-programmers.

(a)

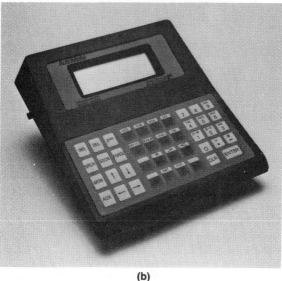

(b)

**Figure 6-4. (a)** Texas Instruments' mini-programmer (non-smart), for use with the model 510 PC, and **(b)**, Reliance Electric's mini-programmer (micro-based), for use with the AutoMate 15 PC.

## Program Loaders

As the name implies, program loaders are used primarily in the loading or reloading of the control program into the programmable controller memory. Once the logic program has been created, edited, and debugged in a CRT or manual programmer, it is normally transmitted to the loader, either directly or via the PC, and stored for later retrieval. There are essentially two types of program loaders: cassette recorders and electronic memory modules.

**Cassette Recorders.** These devices have been used throughout the industry for many years as a means of storing and retrieving control programs. Figure 6-5 shows the STR-LINK, which is one of the most popular digital cassette recorders used with PCs. It communicates to the controller via a standard RS-232 or 20 mA current loop communication interface at speeds that range from 300 to 9600 baud.

The portable audio tape deck is used by some controllers as a means of program storage. Although not as popular as the STR-LINK, these tape decks do provide an inexpensive alternative for storing the program. The Westinghouse manual programmer NLPL-789 is an example of a device that can utilize a standard audio tape deck for program storage (shown in Fig. 6-6).

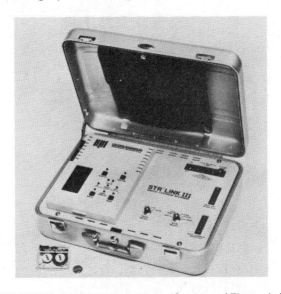

**Figure 6-5.** STAR-LINK III digital cassette recorder (Courtesy of Electronic Processors Inc.).

**Electronic Memory Modules.** These modules are perhaps the smallest storing and reloading devices currently available. They typically contain an EPROM or EEPROM memory with the necessary electronics to write or read a complete PC program into and from the module. Figure 6-7 shows a typical electronic memory module.

Electronic memory modules are inexpensive devices that provide great flexibility for users requiring fast loading of identical programs for several machines. They also allow plant personnel to perform easy loading of field modifications without having to carry bulky programming hardware.

Electronic memory modules are normally available for the low end (small) controllers; however, there is a trend to design larger PCs with this option. Technological improvements in the electronic memory modules will likely allow programs to be stored and retrieved directly from an intelligent CRT or mini-programmer. This improvement will eliminate the need for always having to connect the memory module to the controller.

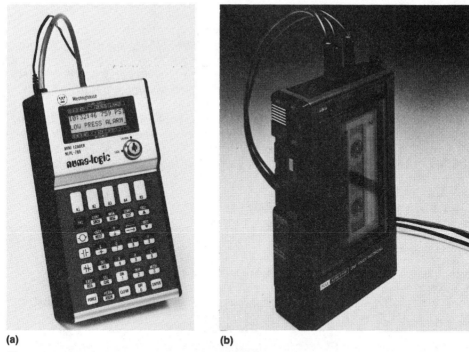

**(a)**                                  **(b)**

**Figure 6-6.** Manual programmer **(a)** and audio cassette recorder **(b)** that can be used to store PC programs (Courtesy of Westinghouse Electric Co.).

**Figure 6-7.** Electronic memory module used with the Modicon Micro-84 PC (Courtesy Gould-Modicon).

## Memory Burners

In spite of the advances in memory and programming technology, there are still few low-cost programmable controllers that utilize ROM, PROM, or EPROM programmers (memory burners) to enter or edit a permanently stored program. To enter a program into these PCs requires that the program be coded and then "blast" onto the memory chips to store the program. These memory chips are then inserted in the CPU as the application memory.

As discussed in Chapter 4, the contents of a ROM or PROM cannot be changed after programmed. If an error is made or a change is needed, it will require a new memory chip and then reprogramming. If the PC uses EPROM memory and a change is required, the chips are reusable after they have been erased first with an EPROM erase device (ultra-violet light), and the program then re-entered.

## Computers

Some PC manufacturers are providing personal computers with special software designed to allow them to be used as programming devices.

A computer as a programming device can provide several advantages over most conventional programming units. A computer can be used to perform general purpose computing tasks when it is not in use (word processing, financial calculations, etc.). It can also serve (if software is provided) to program several different brands of PCs. The UP/DOC (Universal Programmer/Documentation System) by X-Cel Controls uses a personal computer with special software that allows it to program several major PC brands (shown in Figure 6-8). In addition to providing normal programming functions, this system can also serve as a complete control program documentation system for the PCs that it supports.

**Figure 6-8.** UP/DOC system (Courtesy of X-Cel Controls).

## 6-2   OTHER PERIPHERALS

Several peripheral devices have been developed to improve communication between the programmable control system and the system operators. The information that these peripherals provide allows the operators to respond quickly to critical situations, and to supervise efficiently the controller's operation. They also provide useful information that can be the basis for management and maintenance decisions. These devices fall into the four major categories listed below:

1) Data Entry
2) Documentation and Reporting
3) Displays
4) Control Devices

## Data Entry Devices

**Thumbwheel Switches.** Other than the standard programming devices, thumbwheel switches (TWS) are probably the most common means of entering data into the controller. Thumbwheel switches are typically used to enter/modify preset values for timers, counters, shift registers, or other values, such as high and low limits. The thumbwheel switch can also be used like a selector switch by simply assigning meaning to different numbers and then having the program decode the entered number. The numbers 1 through 10, for example, could call for different machine setups.

Physical size of TWS units vary, but are normally 3 to 6 inches in length, 2 to 4 inches in height, and 2 to 3 inches in depth. Typically, thumbwheel assemblies will have 3 or 4 digits, although they may have as many as 6 digits. Figure 6-9 shows a four digit TWS unit. The TWS digits and common wires are brought out to either a terminal strip or connector at the edge of the assembly. The TWS is interfaced to the PC using four wires for each digit (BCD) and is normally connected to a register input module (See Sect. 5-3).

**Figure 6-9.** A four BCD digit thumbwheel switch (Courtesy of Cincinnati Electrosystems Inc.).

The TWS is normally used in conjunction with a selector switch and/or a pushbutton to tell the PC when to read data or what the data represents. A commonly used configuration allows the selector switch to determine which register the data will be entered into; the pushbutton, when pushed, will then allow the data to be read. With such an arrangement, ten parameters could be altered by the operator using a ten position selector switch and a pushbutton.

**Operator Interface Consoles.** These devices are used as multifunction data entry units (and monitoring if available) that can be used as a simple means of inputting data. Normally, these consoles are in the form of a small enclosed unit with keypad or keypad and display. Operator interface consoles offer an alternative to locating the programming device at the point of data entry. Their function is similar to thumbwheel switches, but include a strobe key that allows the data to be sent to the PC interface on command.

Typically, the console is mounted on the operator's control panel or some other convenient location. The data signals sent to the PC can range from 5 VDC to 24 VDC in BCD format, which makes them compatible with register input modules.

**Figure 6-10.**
Numerical Keyboard entry panel
(Courtesy of Cincinnati Electrosystems Inc.).

The keypad normally uses standard raised keys or tactile feedback membrane keys. Some consoles may incorporate an LED seven-segment display which allows the user to view the data being entered or to monitor register or bit values sent from the controller. Figure 6-10 shows a typical operator console.

## Documentation/Reporting

**Program Documentation.** Basically, these systems solve the problem of interpreting the program by providing printouts that include familiar symbols, symbolic names, description of element functions, diagram commentary, data table values, and rung/line cross references. The detail that these systems provide indeed becomes a powerful and efficient tool for debugging, trouble-shooting, and later understanding the control program. Figure 6-11 illustrates one of the most popular documentation systems available.

**Figure 6-11.** Ladder diagram translator documentation system (Courtesy of Xycom).

The documentation systems are an alternative to documentation done by hand, which typically consumes hundreds of costly hours for preparation and updating. Savings of time and money are easily realized with the use of these systems. More details pertaining to the features and capabilities of documentation systems are covered in Chapter 9.

**Figure 6-12.** Model 43 line printer (Courtesy of Teletype Corp.).

**Line Printer.** A line printer (see Fig. 6-12) is used to provide hard copy printouts of the control program and to generate reports or operator messages. Most programmable controllers communicate with line printers in serial form through a standard RS-232 port that is located in either the CPU or in the programming device. Some controllers may provide a serial communications module to allow interfacing with printers and other peripherals.

Some important considerations when selecting a printer include: the baud rate transmission, which is the rate at which data is transmitted/received between the printer and controller; the printer's buffer size, which will allow data to be sent at faster rates and printed at slower rates; and the characters printed per second (cps), which defines the number of characters that the printer can print in one second. These specifications have a definite impact on the time necessary for either printing a program or generating a report.

**Report Generation.** Report generation systems are used to produce reports of production and system operation by gathering data from PCs and/or information entered at the operator keyboard. These systems provide the formatted report information on a screen display or a hard-copy printout. An important benefit of the report generation system is that the PC is free from doing the report itself. External report generation unburdens the PC's processing time and frees memory storage area. Typical information that is reported includes machine cycle times, part counts, material wait times, machine downtime, quality assurance data, scheduling data, and any other useful information that can be gathered.

One example of a report generation system is the XYCOM 4872-RG software package that is used with XYCOM's 4820 ladder documentation system. This package produces reports by extracting from the PC data stored in binary, BCD, or ASCII format. It then converts the data into legible format before the report is generated. Figure 6-13 shows an example report generation from the XYCOM system.

The XYCOM system will also allow keyboard data entry by the operator, if needed. One advantage of this system is that no computer programming experience is required; the user needs only to define a type of report, and the schedule of reporting, through a menu driven operation.

```
                 FIRTS SHIFT PRODUCTION ANALYSIS REPORT
                                  FOR
                 DATE 02-JUL-83  AND   TIME 15:01:27

                                        RUN    DOWN   IDLE
                PASSED  REJECTED  REWORKED  TIME   TIME   TIME
                ------  --------  --------  ----   ----   ----

    LINE #1     382     53        12        6.9    0.0    1.1

    LINE #2     0       0         0         0.0    0.0    0.0

    LINE #3     393     46        36        7.1    0.1    0.8

    LINE #4     355     63        35        6.5    0.3    1.2

    LINE #5     304     50        18        6.3    0.9    0.8


    NOTE:   LINE #2 DOWN FOR SCHEDULED MAINTENANCE

    OPERATOR'S NAME:   TOM DYOLL
```

**Figure 6-13.** Example of Xycom's report generation.

## Display Devices

**Seven-Segment Display.** Figure 6-14 shows a seven-segment display device. These devices are used to display numerical data that is stored in memory (e.g. volume, distance, velocity, and time). The display segments are generally LEDs or LCDs and are available in groups of 4, 5, or 6 digits. The data received by the seven-segment display is normally BCD and is generally interfaced to the controller through a register output module (See Sect. 5-3).

**Figure 6-14.** Seven segment display assembly (Courtesy Cincinnati Electrosystems Inc.).

**Intelligent Alphanumeric Displays.** These intelligent display units are an economical way of enhancing the way PCs provide information to operators regarding the controlled machine or process. Normally, warnings or alarms to the operator are in the form of non-intelligent indicators such as pilot lights or annunciator panels. These old methods of communicating with the operator do not provide complete information and, depending on the number of warnings, can require large amounts of panel space.

Intelligent alphanumeric devices store canned English messages or warnings to the operator that can be displayed under program control. The display information can be either general operations messages to the operator, warnings of failure, or notification of misoperation of any field device. Programming the canned messages for these units is very simple and requires no special training. The messages are normally programmed either by the user or the manufacturer. Figure 6-

15 shows an intelligent alphanumeric display that is easily programmed using an ASCII keyboard. This display can store up to 128 messages, each one containing up to 20 characters. Each message has a specific binary code that is used to generate the message. The display system prompts the programmer to identify the message number and the associated message. A field of up to 4 characters can be reserved in the message to allow variable quantities to be incorporated into the message. The variable data must be BCD and provided by the PC.

Interfacing between the display device and any programmable controller requires seven lines to decode the possible 128 messages and sixteen lines to provide the BCD data, if this option is used.

**Figure 6-15.** Intelligent alphanumeric display that can be used with any programmable controller (Courtesy of Cincinnati Electrosystems Inc.).

**Colorgraphic Displays.** Colorgraphic displays are used in the process industry to provide operators with alarms, display status, messages, flow diagrams, and other useful process information. These graphic displays communicate with PCs to share the information that will be displayed on the color screen, indicating status of the PC operation. This information display actually provides a window into the functional operation of the process or machine, as well as the PC itself. Colorgraphic systems can even communicate with the PC manufacturer's local area networks throughout the factory floor, greatly enhancing the amount of visual information available on command. An industrial colorgraphic display is shown in Fig. 6-16.

**Figure 6-16.** A colorgraphic display system (Courtesy of Industrial Data Terminals Corp.).

## Control Devices

**Manual Control Stations.** Manual control stations (MCS) are devices that provide a manual back-up or manual override of analog or digital controlled devices. Two of the most commonly used control stations are from Control Technology Inc.-- the MCS-7200 and MCS-7408 (Fig. 6-17), which are used for analog and digital back-up control, respectively.

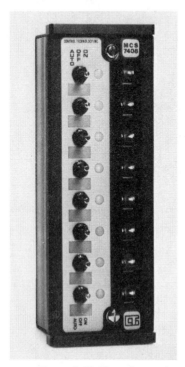
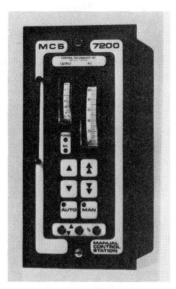

**Figure 6-17.** Manual control stations for **(a)** discrete and **(b)** analog (Courtesy Control Technology Corp.).

The MCS-7200 is used in most cases between the PC's analog interface and some controlled element to provide a means of assuring continued production or control in a variety of abnormal process situations. This control station becomes very useful during start-up, override of analog outputs, and back-up of analog outputs in case of failures.

During start-up, the operator can manually position the final control element by using the MCS-7200 to manipulate initial control parameters, such as valve position, speed control, hydraulic servos, and pneumatic converters. This task can even be accomplished without the presence of the PC or prior to its checkout. When the final elements are working properly, final check with the PC can be accomplished, and the control station can be switched to automatic mode for direct control of the process from the PC.

The MCS-7200 can also be used in applications that require signal level modifications of some combination of analog outputs, such as the addition of more or less ingredients in a batching process. This control station is very helpful in the event of any PC component failure, in which the station could maintain the correct analog output level during the PC failure. After the process is stabilized with the MCS-7200 in the manual mode, the PC failure can be diagnosed and corrective measures can be taken to put the system back in automatic mode. Figure 6-18 illustrates a typical configuration of the use of the MCS-7200 as a back-up unit.

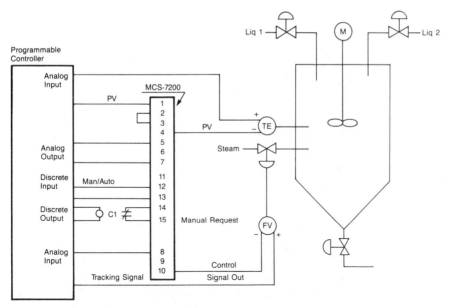

**Figure 6-18.** Typical control station configuration for the MCS 7200 model (Courtesy of Control Technology Corp.).

The MCS-7408 discrete manual back-up station provides eight isolated and circuit breaker points for use with any PC's discrete output modules. This control station is used between the PC's output interface and the digital controlled element. The primary use of the MCS-7408 is to provide a means of controlling a field device with or without PC control. Using this station, the device can be disabled (OFF), or it can be activated (ON) without the PC being in control. Figure 6-19 shows a simplified circuit of the MCS-7408 and typical load connection.

The uses of this control station prove to be very helpful during emergency disconnect of particular field devices, maintenance situations, and especially during the system start-up. Indicators are also incorporated into the control station to show the ON or OFF state of the field device.

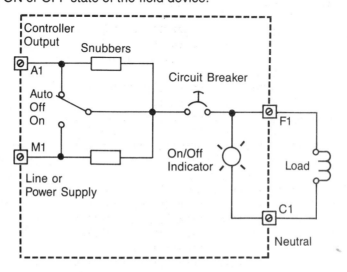

**Figure 6-19.** Typical control station configuration for the MCS 7408 model (Courtesy of Control Technology Corp.).

**Figure 6-20.** Real time clock module (Control Technology Corp.).

**Real-Time Clock Modules.** The real-time clock modules are used to provide the PC with information on the year, day of week, month, day of month, hours, minutes, and seconds. This information is kept updated in the module even in the event of a power failure. Figure 6-20 shows a real-time clock module from one of the Control Technology Inc. clock interfaces.

Typical applications of clock modules are found in energy management, process synchronization, time-keeping during power failure, accurate rate calculations, calendar scheduling, and report generation. One advantage of CTI's clock module is that it responds directly to previously defined PC instructions to give the user the ability to set and read the date and time without a sophisticated software program. The initial setting of time and date can be done easily through the front of the module when a jumper in the rear of the module is in the enable front panel position. Time is also shown in an LCD display at the front of the module for easy reading.

## 6-3 PERIPHERAL INTERFACING

Regardless of the peripheral used, the user must properly connect the device to the PC, or smart module, to achieve correct communication. Typical peripherals communicate in serial form at speeds ranging from 110 to 19,200 bits per second (baud), with parity or non-parity, asynchronously, and by using different communication interface standards.

### Communication Standards

Communication standards fall into two categories: *proclaimed* and *de facto*. Proclaimed standards have been officially established by various electronic organizations, such as the Institute of Electrical and Electronics Engineers (IEEE) or the Electronic Industries Association (EIA). These institutions attempt to define public specifications, by which manufacturers can properly establish communication schemes that allow compatibility among different manufacturers' products. Proclaimed standards, such as the IEEE 488 instrument bus, the EIA RS-232C, and the EIA RS-422, are examples of well-defined standards.

When we refer to de facto standards, we are referring to interface methods that have been adopted and have gained popularity through widespread use without official definition. Some de facto standards have caused interfacing problems in the past; however, other standards, such as the PDP-11 Unibus and 20 mA current loop, are good examples of well-defined de facto standards.

**Serial Communication.** This type of communication, as the name implies, is done in a serial form through simple twisted pair cables. ASCII information is usually sent to peripheral equipment such as terminals, modems, and line printers. Serial data transmission is used for most peripheral communication, since these devices are slow-speed in nature and require long cable connections.

Two of the most popular standards for serial communications are the RS-232C and the 20 ma current loop. A more recent standard is the RS-422, which will tend to improve performance and give greater flexibility in data communication interfaces.

Data communication links generally used with peripheral equipment can be *unidirectional* or *bidirectional.* If a peripheral is strictly either an input or an output device, data needs to be sent only in one direction. In either case, a serial signal line is all that is required to complete the link. Devices that serve as both input and output devices (e.g. video terminal) require bidirectional links. There are two ways to achieve this bidirectional communication. First, a single data line can be provided as a shared communication line. The data can be sent in either direction, but only in one direction at a time. This operation is known as *half-duplex.* If simultaneous bidirectional communication is required, two lines can be sent from the PC to the peripheral. One line would be assigned permanently to an input, while the other would be an output. This mode is known as *full-duplex.* Figure 6-21 illustrates the unidirectional, full-duplex, and half-duplex communication methods.

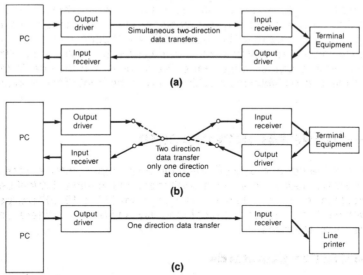

**Figure 6-21.** Data communication formats. **(a)** Full-duplex. **(b)** Half-duplex. **(c)** Unidirectional.

**EIA RS-232C.** The EIA RS-232C is a proclaimed standard that defines the interfacing between data equipment and communication equipment employing serial binary data interchange. Electrical signals, as well as mechanical details of the interface, are well-defined by this standard. The complete RS-232C interface consists of 25 data lines, which have all possible signals for a simple to complex communication interface. Although several of these lines are specialized and few are left undefined, most peripherals require only from 3 to 5 lines to operate properly. Table 6-1 describes the 25 lines as specified by EIA.

**Table 6-1.** EIA RS-232C Interface Signals

| Pin Number | Description |
|---|---|
| 1 | Protective Ground |
| 2 | Transmitted Data |
| 3 | Received Data |
| 4 | Request to Send |
| 5 | Clear to Send |
| 6 | Data Set Ready |
| 7 | Signal Ground (Common Return) |
| 8 | Received Line Signal Detector |
| 9 | (Reserved for Data Set Testing) |
| 10 | (Reserved for Data Set Testing) |
| 11 | Unassigned |
| 12 | Secondary Received Line Signal Detector |
| 13 | Secondary Clear to Send |
| 14 | Secondary Transmitted Data |
| 15 | Transmission Signal Element Timing (DCE) |
| 16 | Secondary Received Data |
| 17 | Receiver Signal Element Timing (DCE) |
| 18 | Unassigned |
| 19 | Secondary Request to Send |
| 20 | Data Terminal Ready |
| 21 | Signal Quality Detector |
| 22 | Ring Indicator |
| 23 | Data Signal Rate Selector (DTE/DCE) |
| 24 | Transmit Signal Element Timing (DTE) |
| 25 | Unassigned |

Figure 6-22a illustrates an RS-232C data communication system using a telephone modem (modulator/demodulator); 6-22b shows the RS-232C wiring connections from a computer to a smart EIA PC interface module; and 6-22c illustrates a typical interfacing to a teletype printer.

Note that the communication between a computer and a PC has few lines swapped, if no modem or other data communication equipment is used. This cable is called a *null modem* cable. When connecting a PC to an RS-232C peripheral (printer, etc.), four wires are normally required; however, it is recommended that users refer to the connection specifications for both devices for any special details.

The RS-232C standard calls for certain specific electrical characteristics. A listing of some of these specifications follows:

a)   The signal voltages at interface point are a minimum of +5V and maximum of +15V for logic 0, and for logic 1, a minimum of -5V and maximum of -15V.

b)   The maximum recommended distance is 50 feet or 15 meters; however, longer distances are permissible, provided that the resulting load capacitance measured at the interface point and including the signal terminator, does not exceed 2500 picofarads.

c)   The drivers used must be able to withstand open or short circuits between pins in the interface.

d)   The load impedance at the terminator side must be between 3000 and 7000 ohms, with no more than 2500 picofarads capacitance.

e)   Voltages under -3V (logic 1) are called MARK potentials (signal condition); voltages above +3V (logic 0) are called SPACE voltages. The area between -3V and +3V is not defined.

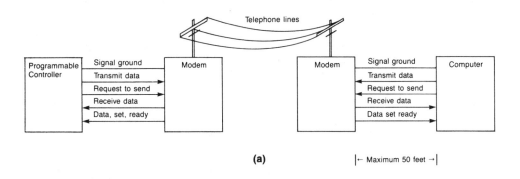

**(a)**

$|\leftarrow$ Maximum 50 feet $\rightarrow|$

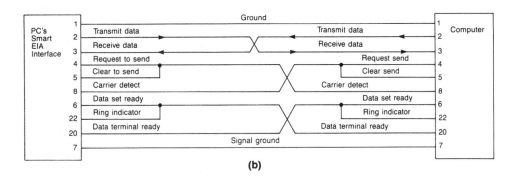

**(b)**

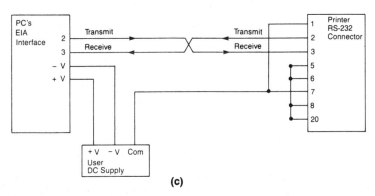

**(c)**

**Figure 6-22.** RS-232C communication connections for **(a)** modem, **(b)** PC to computer, and **(c)** PC to printer.

Figure 6-23 illustrates a typical RS-232C serial ASCII pulse train. The transmission starts with a *start* bit (0) and ends with either one or two *stop* (1s) bits. Parity is also included and can be even or odd (see Chapter 4 for parity).

**EIA RS-422.** The RS-422 standard has been devised to overcome some of RS-232C shortcomings, including an upper data-rate of 20 Kbaud, maximum cable distance of 50 feet, and an insufficient capacity to control additional loop-test functions for fault isolation. The RS-422 standard still deals with the traditional serial-binary switch-signals of two voltage levels across the interface. Mechanical specifications for this electrical interface standard (RS-422) are defined by the RS-449 standard to meet new operational requirements.

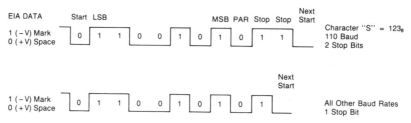

**Figure 6-23.** RS-232C serial ASCII pulse train.

The RS-232C is an *unbalanced link* communication which specifies a primary station always in control (Master/Slave) and responsible for setting logical states and operational modes of each secondary station, thereby controlling the entire data communication process. The RS-422 is a *balanced link* in which either party can configure itself and initiate transmission, when both stations have identical data transfer and link control capability. The RS-422 specifies electrically balanced receivers and generators that tolerate and produce less noise to provide superior performance up to 10 Megabaud (10,000 Kbaud) and to meet even more requirements in the future.

The balanced circuit employs differential signaling over a pair of wires for each circuit. The unbalanced configuration signals use one wire for each circuit and a common return circuit. Figure 6-24 illustrates both configurations for RS-422 and RS-232C.

In general, the RS-422 may be required when interconnecting cables are too long for effective unbalanced operation, and noise in excess of 1 volt can be measured across the signal conductors. The driver circuits for RS-422 have the capability to furnish the DC signal necessary to drive up to 10 parallel connected RS-422 receivers; however, this capability involves considerations such as stub line lengths, data rate, grounding, fail-safe networks, etc. Cable characteristics for RS-422 are not specified, but to insure proper operation, paired cables with metallic conductors should be employed and, if necessary, shielded.

The maximum distance applicable for RS-422 is a function of the data transmission rate. A relationship between distance and data rate is illustrated in Fig. 6-25. The balanced electrical characteristics of RS-422 provide an even better performance with an optimal cable termination of approximately 120 ohms in the receiver load. These curves, however, are conservative for RS-422 balanced operation. Actually, several miles can be realized at lower data rates with good engineering practice. The graph here describes empirical measures using a 24 AWG, copper conductor, twisted pair cable with a shunt capacitance of 52.5 pF/meter (16 pF/foot), terminated in a 100 ohm resistive load. If longer distances are required, analysis on absolute loop resistance and capacitance of the cable should be performed. Longer distances, in general, would be possible when using 19 AWG cable. The type and length cable used must be capable of maintaining the necessary signal quality needed for the particular application.

The RS-449 mechanical standard (that supports the electrical RS-422) offers several extra circuits (signals) that were added to provide greater flexibility to the interface and to accommodate the new common return circuits. These additional functions and wires were beyond the capacity of the RS-232C 25-pin connector; therefore, EIA selected a 37-pin connector that will satisfy the needs of the interface channels. If secondary channel operation is to be used as a low-speed TTY or acknowledgements channel, a separate nine-pin connector is also needed.

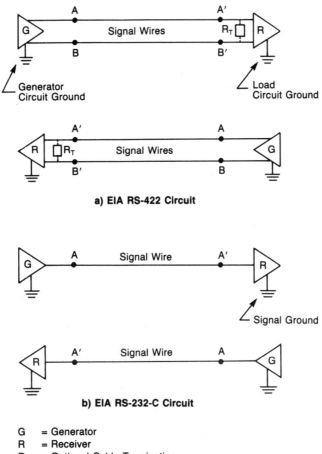

**a) EIA RS-422 Circuit**

**b) EIA RS-232-C Circuit**

G   = Generator
R   = Receiver
$R_T$  = Optional Cable Termination
A,B,A',B' = Interface Points

**Figure 6-24.** Configuration of (a) RS-422 and (b) RS-232C.

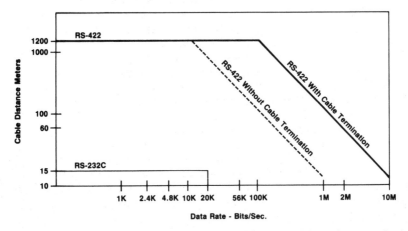

**Figure 6-25.** Data rate and distance relationship for RS-422 and RS-232C communication standards.

**20 mA Current Loop.** The 20 mA current loop de facto standard consists of four basic wires: transmit plus, transmit minus, receive plus, and receive minus. Figure 6-27 illustrates the four lines used to form the 20 mA current loop.

Recognition of ones and zeroes is achieved by opening and closing the current loop. When the current loop standard was first used in teletypewriters, the loop was connected and broken by rotating switch contacts within the teletypewriter sending the data, and the 20 mA signal drove a print magnet in the receiving teletypewriter. Today, most 20 mA current loops electronically operate the opening switch and printer magnet arrangement.

The voltage in the 20 mA current loop is applied to a current limiting resistor at the data sending end, thus generating the current. The voltage is dropped across current limiting resistor $R_{TX}$ and also across the load resistor $R_L$. The values of $R_{TX}$ and the positive voltage applied to it must generate a flow current of 20 mA. Typically, a high voltage and high value resistance ($R_{TX}$) are usually chosen, even though a low voltage and low resistance could be used. Current loop communications present a big advantage, since the wire resistance has no effect on a constant current loop. Voltage does not drop across the wire as it does in the RS-232C voltage-oriented interface, thus allowing the current loop interface to drive signals longer distances. To achieve this advantage, a constant current source is generally used to generate the 20 mA current.

Converting 20 mA current loop to RS-232C can be done simply by employing an RS-232C level receiver to drive a switching transistor on the transmission end and an optical-isolator and load resistor to drive the RS-232C driver on the receiving end.

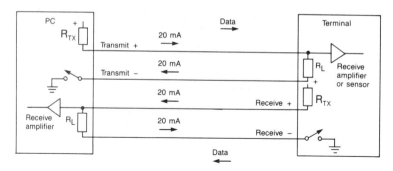

**Figure 6-26.** Illustration of 20 mA current loop operation.

# PART
# --]III[--

# UNDERSTANDING THE SOFTWARE COMPONENTS

This part surveys the common programming languages, describes the various instructions, presents the approach to program design, offers several programming techniques and examples, and reviews the elements of a good documentation package. Although the chapters in this part are independent, readers may appreciate them more if they are read sequentially.

# CHAPTER
# --]7[--

# PROGRAMMING
# LANGUAGES

*The chief merit of language is clearness.*
—Galen

This chapter will present the various languages currently used to create the programmable controller program. It will describe the most commonly found instructions used to implement each of these languages. In general, regardless of the particular language used, the same basic functions are achieved.

# 7-1  INTRODUCTION

There are four types of languages normally encountered in programmable controllers:

- Ladder Diagrams
- Functional Blocks
- Boolean Mnemonics
- English Statements

These languages can be grouped into two major categories. The first two, ladder and Boolean, form *basic PC languages*, while functional blocks and English statements are considered *high level languages*. The basic PC languages consist of a set of instructions that will perform the most primitive type of control functions: relay replacement, timing, counting, sequencing, and logic. However, depending on the controller model, the instruction set may be extended or enhanced to perform other basic operations. The high level languages have been brought about by a need to execute more powerful instructions that go beyond the simple timing, counting, and ON/OFF control. High level languages are suited for operations such as analog control, data manipulation, reporting, and other functions that are not possible with the basic instruction sets.

The language used in a PC actually dictates the range of applications in which the controller can be applied. Depending on the size and capabilities of the controller, one or more languages may be used. Typical combinations of languages are:

- Ladder Diagrams only
- Boolean only
- Ladder Diagrams and Functional Blocks
- Ladder Diagrams and English Statements
- Boolean and Functional BLocks
- Boolean and English Statements

# 7-2  PC INSTRUCTION SUMMARY

This section contains an overview of the various programmable controller instructions. These instructions form a set of tools that include all the machine functions to perform the following six operations:

- Relay Logic
- Arithmetic
- Data Transfers
- Timing and Counting
- Data Manipulation
- Flow of Control

The ladder and Boolean languages both include, as a basic set, the relay logic as well as timer and counter instructions. These two operations are normally found in small controllers. Further enhancements in these languages will include one or more of the remaining types of operations. The functional block and English statement languages typically include all of these operations; however, the high level language instruction sets will include more instructions, with enhancements to any instructions that are commonly found in either ladder or Boolean. Table 7-1 summarizes the six major types of operations and the instructions that are found under these categories.

**Table 7-1.** PC instruction summary.

| Operation Type | Basic Level Language | High Level Language |
|---|---|---|
| Relay Logic | ‖ ‖/ ‖ ( ) (/) (L) (U) ‖↑‖ ‖↓‖ | |
| Timer and Counter | (TON) (TOF) (RTO) (RTR) (CTU) (CTD) (CTR) | TMR On<br>TMR Off<br>Count Up<br>Count Down<br>Count Up/Down |
| Arithmetic | ‖+‖ ‖-‖ ‖ ‖ ‖÷‖ | Add<br>Sub<br>Mul<br>Div<br>Double Prec Add<br>Double Prec Sub<br>Double Prec Mul<br>Double Prec Div<br>Square Root<br>Floating Point<br>Trigonometric |

**Table 7-1 (continued).** PC instruction summary.

| | | |
|---|---|---|
| Data Manipulation | —\|CMP =\|—  —\|CMP >\|—  —\|CMP <\|— | CMP = <br> CMP > <br> CMP < <br> CMP ≥ <br> CMP ≤ <br> LIMIT <br> Logic Matrix <br> Convert BCD → BIN <br> Convert BIN → BCD <br> Absolute <br> Invert <br> Complement <br> Set Constant <br> Shift <br> Rotate <br> Examine Bit <br> Increment Register |
| Data Transfer | —\| GET \|—  —(PUT)— | Move Register <br> Move Point <br> Move With Mask <br> Move Block <br> Table To Register <br> Register To Table <br> Block Transfer In <br> Block Transfer Out <br> ASCII Transfer <br> FIFO Stack Transfer |
| Program Flow Control | —(MCR)—  —(ZCL)—  —(SKD)—  —(SKR)—  —(JMP)—  —\|LBL\|—  —(JSB)—  —(RET)— | |

## 7-3  LADDER DIAGRAM LANGUAGE

The ladder diagram language is a symbolic instruction set that is used to create a programmable controller program. It is composed of five categories of instructions that include *relay-type, timer/counter, arithmetic, data manipulation, data transfer, and program control*. The ladder instruction symbols can be formatted to obtain the desired control logic that is to be entered into memory. Because the instruction set is composed of contact symbols, it is also referred to as *contact symbology*.

The main function of the ladder diagram program is to control outputs based on input conditions. This control is accomplished through the use of what is referred to as a *ladder rung*. Figure 7-1 shows the basic structure of a ladder rung. In general, a rung consists of a set of input conditions, represented by contact instructions, and an output instruction at the end of the rung, represented by the coil symbol. Throughout this section, the contact instructions for a rung may be referred to as input conditions, rung conditions, or control logic.

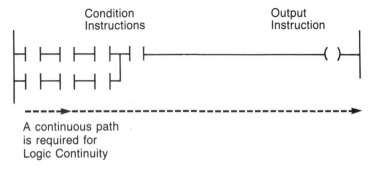

**Figure 7-1.** Ladder diagram rung format.

### Basic Symbols

Coils and contacts are the basic symbols of the ladder diagram instruction set. The contact symbols programmed in a given rung represent conditions to be evaluated in order to determine the control of the output; all outputs are represented by coil symbols.

When programmed, each contact and coil is referenced with an address number which identifies what is being evaluated and what is being controlled. Recall that these address numbers reference the data table location of either an internal (storage bit) output or a connected input or output. A contact, regardless of whether it represents an input/output connection or an internal output, can be used throughout the program whenever that condition needs to be evaluated.

The format of the rung contacts is dependent on the desired control logic. Contacts may be placed in whatever *series, parallel,* or *series/parallel* configuration is required to control a given output. For an output to be activated or energized, at least one left-to-right path of contacts must be closed. A complete closed path is referred to as having *logic continuity*. When logic continuity exists in at least one path, it is said that the rung condition is TRUE. The rung condition is FALSE if no path has continuity.

For several years, the standard ladder instruction set was limited to performing only relay equivalent functions, using the basic relay-type contact and coil symbols similar to those illustrated in Fig. 7-1. A need for greater flexibility, coupled with developments in technology, has led to extended ladder diagram instruction sets that perform data manipulation, arithmetic, and program flow control. It is now commonplace to find controllers that include computer-like expressions similar to the diagram in Fig. 7-2. Further enhancements to the extended ladder diagram instruction set have resulted in functional block instructions.

Although instructions and symbols may differ among controllers, the instructions described here are generic and apply to most controllers.

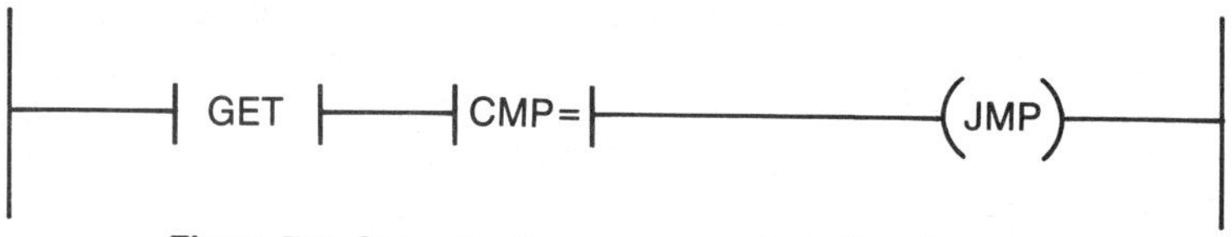

**Figure 7-2.** Computer like expressions in ladder diagram format.

## Relay-Type Instructions

The relay-type instructions are the most basic of programmable controllers instructions. They provide the same capabilities as hardwired relay logic, but with greater flexibility. These instructions primarily provide the ability to examine the ON/OFF status of specific bit addresses in memory and to control the state of an internal or external output. The following is a description of relay-type instructions that are most commonly available in any controller that has a ladder diagram instruction set.

**Symbol: --]  [--**
**Name:** Normally-Open Contact

The normally-open contact is programmed when the presence of the referenced signal is needed to turn an output ON. When evaluated, the referenced address is examined for an ON (1) condition. The referenced address may represent the status of an external input, external output, or internal output. If, when examined, the referenced address is ON, then the normally-open contact will close and allow logic continuity (power flow). If it is OFF (0), then the normally-open contact will assume its normal programmed state (open), thus breaking logic continuity.

**Symbol: --]/[--**
**Name:** Normally-Closed Contact

The normally-closed contact is programmed when the absence of the referenced signal is needed to turn an output ON (1). When evaluated, the referenced address is examined for an OFF (0) condition. The referenced address may represent the status of an external input, external output, or an internal output. If, when examined, the referenced address is OFF, then the normally-closed contact will remain closed allowing logic continuity. If the referenced address is ON, then the normally-closed contact will open and break logic continuity.

**Symbol: --(  )--**
**Name:** Energize Coil

The energize coil instruction is programmed to control either an output connected to the controller or an internal (control relay) output. If any rung path has logic continuity, the referenced output is turned ON. The output is turned OFF if logic continuity is lost. When the output is ON, a normally-open contact of the same address will close, and a normally-closed contact will open. If the output goes OFF, any normally-open contact will then open, and normally-closed contacts will close. An example rung using the previous instructions is shown in Fig. 7-3.

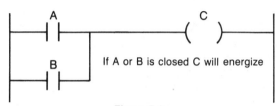

**Figure 7-3.**

**Symbol: --(/)--**
**Name:** De-energize Coil

The de-energize coil instruction is programmed to control either an internal output or an output device connected to the controller. If no rung path has logic continuity, the referenced output is turned ON. The output is turned OFF if logic continuity is achieved. When the output is ON, normally-open contacts of the same address will be closed, and normally-closed contacts will be open. If the output goes OFF, normally-open contacts will open, and normally-closed contacts will close.

**Symbol: --(L)--**
**Name:** Latch Coil

The latch coil instruction is programmed if it is necessary for an output to remain energized, even though the status of the contacts which caused the output to energize may change. If any rung path has logic continuity, the output is turned ON and retained ON, even if logic continuity or system power is lost. The latched output will remain latched ON until it is unlatched by an unlatch output instruction of the same reference address. The unlatch instruction is the only automatic (programmed) means of resetting the latched output. Although most controllers allow latching of internal or external outputs, some are restricted to latching internal outputs only.

**Symbol: --(U)--**
**Name:** Unlatch Coil

The unlatch coil instruction is programmed to reset a latched output of the same reference address. If any rung path has logic continuity, the referenced address is turned OFF. The Unlatch output is the only automatic means of resetting a latched output. Figure 7-4 illustrates the use of the latch and unlatch coils.

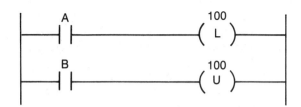

If A closes, coil 100 is latched until B closes to unlatch coil 100.

**Figure 7-4.** Example of latch and unlatch coil instructions.

**Symbol: --]↑[--**
**Name:** OFF-ON Transitional Contact

The OFF-ON transitional contact is programmed to provide a one-shot pulse when the referenced trigger signal makes a positive (OFF-to-ON) transition. This contact will close for exactly one program scan whenever the trigger signal goes from OFF-to-ON. The contact will allow logic continuity for one scan and then open, even though the triggering signal may stay ON. The triggering signal must go OFF and ON again for the transitional contact to close again. The contact address (trigger) may represent an external input/output or an internal output. Operation of the OFF-to-ON transitional is illustrated in Fig. 7-5.

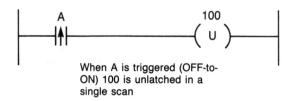

When A is triggered (OFF-to-ON) 100 is unlatched in a single scan

**Figure 7-5.**

**Symbol: --]↓[--**
**Name:** ON-OFF Transitional Contact

The ON-OFF transitional contact is programmed to provide a one-shot pulse when the referenced trigger signal makes a negative (ON-to-OFF) transition. This contact will close for exactly one program scan whenever the trigger signal makes an ON-to-OFF transition. The contact will allow logic continuity for one scan and then open, even though the triggering signal may stay OFF. The triggering signal must go from OFF to ON and OFF again for the transitional contact to close again. The contact address (trigger) may represent an external input/output or internal output.

## Timer And Counter Instructions

Timers and counters are output instructions that provide the same functions as would hardware timers and counters. They are used to activate or de-activate a device after an expired interval or count. The timer and counter instructions are generally considered internal outputs. Like the relay-type instructions, timer and counter instructions are fundamental to the ladder diagram instruction set.

The operations of the software timer and counter are quite similar, in that they are both counters. A timer counts the number of times that a fixed interval of time (e.g. 0.1 sec, 1.0 sec) elapses. To time an interval of 3 seconds, a timer counts three 1 second intervals (time base).  A counter simply counts the occurrence of an event. Both the timer and counter instructions require an accumulator register (word location) to store the elapsed count and a preset register to store a preset value. The preset value will determine the number of event occurrences or time base intervals that are to be counted.

**Symbol: --(TON)--**
**Name:** Time Delay Energize (ON)

The delay energize timer output instruction is programmed to provide time delayed action or to measure the duration for which some event is occurring. If any rung path has logic continuity (see Fig. 7-6), the timer begins counting time-based intervals and times until the accumulated time equals the preset value. When the accumulated time equals the preset time, the output is energized, and the timed out contact associated with the output is closed. The timed contact can be used

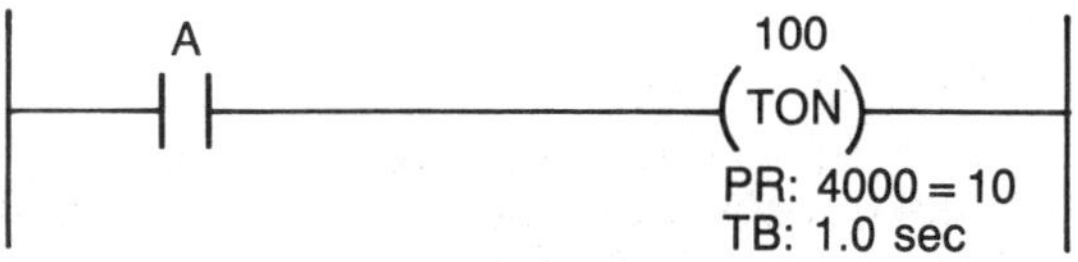

Timer 100 is energized 10
seconds after A closes.

**Figure 7-6.**

throughout the program as a NO or NC contact. If logic continuity is lost before the timer is timed out, the accumulator register is reset to zero.

### Symbol: --(TOF)--
**Name:** Time Delay De-energize (OFF)

The delay de-energize timer output instruction is programmed to provide time delayed action. If logic continuity is lost, the timer begins counting time-based intervals until the accumulated time equals the programmed preset value. When the accumulated time equals the preset time, the output is de-energized, and the timed-out contact associated with the output is opened. The timed contact can be used throughout the program as a NO or NC contact. If logic continuity is gained before the timer is timed out, the accumulator is reset to zero.

### Symbol: --(RTO)--
**Name:** Retentive ON-Delay Timer

The retentive timer output instruction is programmed if it is necessary for the timer accumulated value to be retained, even if logic continuity or power is lost. If any rung path has logic continuity, the timer begins counting time-base intervals until the accumulated time equals the preset value. The accumulator register retains the accumulated value, even if logic continuity is lost before the timer is timed out, or if power is lost. When the accumulated time equals the preset time, the output is energized, and the timed out contact associated with the output is turned ON. The timer contacts can be used throughout the program as a NO or NC contact. The retentive timer accumulator value must be reset by the retentive timer reset instruction.

### Symbol: --(RTR)--
**Name:** Retentive Timer Reset

The retentive timer reset output instruction is the only automatic means of resetting the accumulated value of a retentive timer. If any rung path has logic continuity, then the accumulated value of the referenced retentive timer is reset to zero.

### Symbol: --(CTU)--
**Name:** Up-counter

The up-counter output instruction will increment by one each time the counted event occurs. A control application of a counter is to turn a device ON or OFF after reaching a certain count. An accounting application of a counter is to keep track of the number of filled bottles that pass a certain point. The up-counter increments its accumulated value each time the up-count event makes an OFF-to-ON transition. When the accumulated value reaches the preset value, the output is turned ON, and the count finished contact associated with the referenced output is closed. Depending on the controller, after the counter reaches the preset value, the counter is either reset to zero or continues to increment for each OFF-to-ON transition. If the latter case is TRUE, a reset instruction is used to clear the accumulator.

137

**Symbol: --(CTD)--**
**Name:** Down-counter

The down-counter output instruction will count down by one each time a certain event occurs. Each time the down-count event occurs, the accumulated value is decremented. In normal use, the down-counter is used in conjunction with the up-counter to form an up/down counter. In this application, the down-counter provides a means for accounting data correction. For example, while the CTU counts the number of filled bottles that pass a certain point, a CTD with the same reference address would subtract one from the accumulator each time an empty or improperly filled bottle occurs. Depending on the controller, the down-counter will stop at zero or at the maximum negative value.

**Symbol: --(CTR)--**
**Name:** Counter Reset

The counter reset output instruction is used to reset the CTU and CTD accumulated values. When programmed, the CTR coil is given the same reference address as the CTU and CTD coils. If the CTR rung condition is TRUE, the referenced address will be cleared.

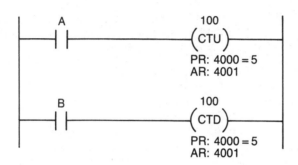

Counter 100 increments by one when A closes, and decrements by one if B closes. The counter energizes after the count reaches 5.

# Arithmetic Operations

The arithmetic operations include the four basic operations of addition, subtraction, multiplication, and division. These instructions use the contents of two registers and perform the desired function.

**Symbol: --( + )--**
**Name:** ADD

The ADD instruction performs the addition of two values stored in the referenced memory locations. How these values are accessed is dependent on the controller. Some instruction sets use a GET (data transfer) instruction to access the two operand (registers) values, while others simply reference the two registers using contact symbols. The result is stored in the register referenced by the add coil. If the addition operation is enabled only when certain rung conditions are TRUE, then the input conditions should be programmed before the values are accessed in the addition rung. Overflow conditions are usually signalled by one of the bits of the addition result register.

**Symbol: --(-)--**
**Name:** SUB

The SUB instruction performs the subtraction operation of two registers. As in addition, if there is a condition to enable the subtraction, it should be programmed before the values are accessed in the rung. The subtraction result register will usually have an underflow bit to represent a negative result.

**Symbol: --(X)--(X)--**
**Name:** MUL

The MUL instruction performs the multiplication operation. It uses two registers to hold the result of the operation between two operand registers. The two registers are referenced by two output coils. If there is a condition to enable the operation, it should be programmed before the two operands are accessed in the multiplication rung.

**Symbol: --( ÷ )--( ÷ )--**
**Name:** DIV

The DIV instruction performs the quotient calculation of two numbers. The result of the division is held in two result registers as referenced by the output coils. The first result register generally holds the integer, while the second result register holds the decimal fraction.

## Data Manipulation Operations

The data manipulation instructions are an enhancement of the basic ladder diagram instruction set. Whereas the relay-type instructions were limited to the control of internal and external outputs based on the the status of specific bit addresses, the data manipulation instructions allow multi-bit operations. In general, the manipulation of data using ladder diagram instructions involves simple register (word) operations to compare the contents of two registers. In the ladder language, there are three basic data manipulation instructions: *compare equal to, compare greater than,* and *compare less than.* Based on the result of a greater than, less than, or equal to comparison, an output can be turned ON or OFF, or some other operation can be performed.

**Symbol: --]CMP = [--**
**Name:** Compare Equal

The compare equal instruction is used to compare the contents of two referenced registers for an equal condition, when the rung conditions are TRUE. As in the arithmetic instructions, the values to be compared can be accessed directly or through a GET instruction depending on the controller. If the operation is TRUE, the output coil is energized.

**Symbol: --]CMP < [--**
**Name:** Compare Less Than

Similar to the compare equal, the less than comparison tests the contents of the value of one register to see if it is less than the value stored in a second register. If the test condition is TRUE, the output coil is energized.

**Symbol: --]CMP > [--**
**Name:** Compare Greater Than

The compare greater than operates like the compare less than, with the exception that the test is performed for a greater than condition. If the test condition is TRUE, the output coil is energized.

## Data Transfer Operations

Data transfer instructions simply involve the transfer of the contents from one register to another. Data transfer instructions can address any location in the memory data table, with the exception of areas restricted to user application. Prestored values can be automatically retrieved and placed in any new location. That location may be the preset register for a timer or counter or even an output register that controls a seven-segment display.

**Symbol: --]GET[--**
**Name:** Get Word

The GET instruction accesses the contents of the referenced address and makes it available for other operations. Some controllers use this instruction to access registers to perform math operations or comparisons.

**Symbol: --(PUT)--**
**Name:** Put Word

The PUT instruction is used to store the result of other operations in the memory location (register) specified by the PUT coil. PUT is generally used with the GET instruction to perform a move register operation. Figure 7-7 shows an example ladder diagram using arithmetic, data manipulation, and data transfer instructions.

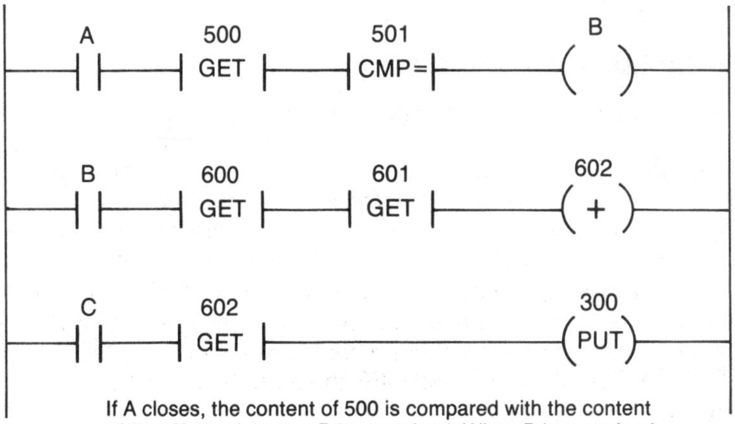

If A closes, the content of 500 is compared with the content of 501. If equal, output B is energized. When B is energized the contents of 600 and 601 are added, and stored in 602. If C closes the content of 602 is transferred to 300.

**Figure 7-7.**

## Program Control Operation

Program control operations are accomplished using a series of conditional and unconditional branches and return instructions. These instructions provide a means of executing sections of the control logic if certain conditions are met.

**Symbol: --(MCR)--**
**Name:** Master Control Relay

The MCR output instruction is used to activate or de-activate the execution of a group (zone) of ladder rungs (see Fig. 7-8). The MCR rung is used in conjunction with an END MCR rung to fence the group of rungs. The MCR rung with conditional inputs is placed at the beginning of the zone, and the END MCR rung with no conditional inputs is placed at the end of the zone. When the MCR rung condition is TRUE, the referenced output is activated, and all rung outputs within the zone can be controlled by their respective input conditions. If the MCR output is turned OFF, all non-retentive (non-latched) outputs within the zone will be de-energized.

**Symbol: --(ZCL)--**
**Name:** Zone Control Last State

The zone control instruction is similar to the MCR instruction. It determines if a group of ladder rungs will be evaluated or not. The ZCL output with conditional

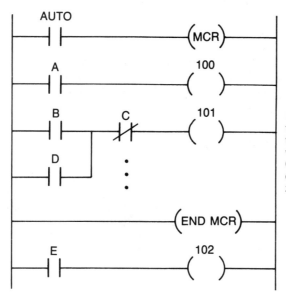

If the AUTO input closes, the MCR is energized and the rungs inside the zone will be executed. If AUTO is OFF, execution resumes at output rung 102 (out of zone).

**Figure 7-8.**

inputs is placed at the start of the fenced zone, and an END ZCL output with no conditional inputs is placed at the end of the zone. If the referenced ZCL output is activated, the outputs within the zone are controlled by their respective rung input conditions. If the ZCL output is turned OFF, the outputs within the zone will be held in their last state.

**Symbol: --(SKD)--**
**Name:** Skip And De-energize
The SKD output instruction functions exactly like the MCR instruction.

**Symbol: --(SKR)--**
**Name:** Skip And Retain
The SKR instruction functions exactly like the ZCL instruction.

**Symbol: --(JMP)--**
**Name:** Jump To Label
The jump instruction allows the normal sequential program execution to be altered if certain conditions exist. If the rung condition is TRUE, the JMP coil reference address tells the processor to jump forward and execute the rung labeled with the same reference address as the JMP coil. In this way, order of execution can be altered to execute a rung that needs immediate attention.

**Symbol: --]LBL[--**
**Name:** Label
The LBL instruction is used to identify a ladder rung which is the target destination of a JMP or JSB (Jump to Subroutine) instruction. The LBL reference number must match that of the JMP and/or JSB instruction with which it is used. The LBL instruction does not contribute to logic continuity, and for all practical purposes is always logically TRUE. It is placed as the first condition instruction in the rung. A LBL instruction referenced by a unique address can be defined only once in a program.

**Symbol: --(JSB)--**
**Name:** Jump to Subroutine

The jump to subroutine output instruction allows the normal program execution to be altered if certain conditions exist. If the rung condition is TRUE, the JSB coil reference address tells the processor to jump to the ladder rung labeled with the same reference number, and continue program execution until a RETurn coil is encountered. Each subroutine begins with a labeled rung and must end with an unconditional RETurn instruction.

**Symbol:   --(RET)--**
**Name:** Return Coil

The return instruction is used only to terminate a ladder subroutine. It is programmed with no conditional inputs. When encountered, program control is returned to the main program and begins at the ladder rung immediately following the JSB instruction that initiated the subroutine. Normal program execution continues from that point. A RET instruction must be programmed for each subroutine. Figure 7-9 illustrates the use of some program control instructions.

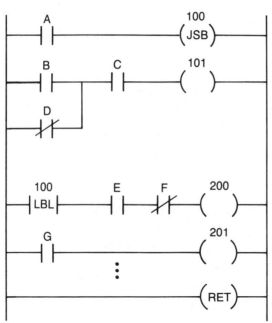

If A closes, the program execution jumps to subroutine label 100. After the subroutine is completed, program execution returns to the rung following the JSB instruction.

**Figure 7-9.**

## Other Ladder Instructions

**Symbol: --(II)--**
**Name:** Immediate Input

The immediate input instruction is used to read an input condition before the I/O update is performed. This operation interrupts the program scan when it is executed. After the immediate input istruction is executed, normal program scan resumes.

**Symbol: --(IO)--**
**Name:** Immediate Output

The immediate output instruction, like the immediate input, updates the specified output immediately.

**Symbol: --(G)--**
**Name:** Global Coil

The global coil is used in a system with a local area network. This output coil operates like any other coil, but its status is sent to all other controllers tied to the network. Each unique global coil address can be defined in only one network controller.

**Symbol: --(GR)--**
**Name:** Global Register

The global register coil operates like the global coil, except that it sends the contents of the referenced coil address to all the controllers in the network.

## 7-4  BOOLEAN LANGUAGE

The Boolean language is a basic level PC language that is based primarily on the Boolean operators: AND, OR, and NOT. A complete Boolean instruction set consists of the Boolean operators and other *mnemonic* instructions that will implement all the functions of the basic ladder diagram instruction set. A mnemonic instruction is written in an abbreviated form, using three or four letters that generally imply the operation of the instruction. Table 7-2 lists a typical set of Boolean instructions and the equivalent ladder diagram symbology.

**Table 7-2.** A Typical Boolean Instruction Set With Extended Functions

| MNEMONIC | FUNCTION | LADDER EQUIVALENT |
|----------|----------|-------------------|
| LD/STR | Load/Start | !--] [-- |
| LD/STR NOT | Load/Start Not | !--]/[-- |
| AND | And Point | --] [-- |
| AND NOT | And Not Point | --]/[-- |
| OR | Or Point | !<br>!<br>--] [--! |
| OR NOT | Or Not Point | !<br>!<br>--]/[--! |
| OUT | Energize Coil | --( )-- |
| OUT NOT | De-Energize Coil | --(/)-- |
| OUT CR | Energize Internal Coil | --( )-- |
| OUT L | Latch Output Coil | --(L)-- |
| OUT U | Unlatch Output Coil | --(U)-- |
| TIM | Timer | --(TON)-- |
| CNT | Up Counter | --(CTU)-- |
| ADD | Addition | --( + )-- |
| SUB | Subtraction | --(-)-- |
| MUL | Multiplication | --(X)-- |
| DIV | Division | --( ÷ )-- |
| CMP | Compare $=, <, >$ | --(CMP)-- |
| JMP | Jump | --(JMP)-- |
| MCR | Master Control Relay | --(MCR)-- |
| END | End MCR, Jump, or Program | --(END)-- |
| ENT | Enter Value for Register | Not Required |

Further enhancements of the Boolean extended instruction sets have resulted in English statement instructions. Figure 7-10 shows a short Boolean program that was converted from a ladder diagram. The principles of Boolean algebra, which are applied in the Boolean language, are discussed in Chapter 3.

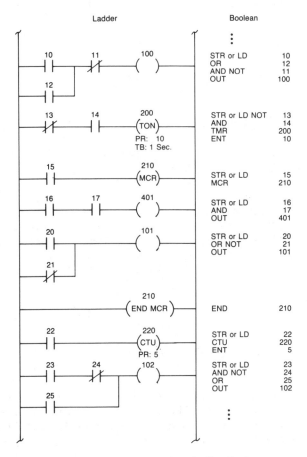

**Figure 7-10.** PC ladder diagram and equivalent Boolean program.

## 7-5 FUNCTIONAL BLOCKS LANGUAGE

Functional blocks are high level instructions that permit the user to program more complex functions using the ladder diagram format. The instruction set is composed of *"blocks"* that execute or perform a specific function.

When using block instructions, input conditions are programmed using NO and NC contacts, which will enable the block operation. There are also several parameters associated with the block that must be programmed. These parameters normally include storage or holding registers used to set preset values, or I/O registers (variables), used to input or output numeric data (analog, BCD, etc.).

Functional blocks, like ladder diagrams, can be classified into four main types: *timer and counter* instructions in block form, *arithmetic, data manipulation,* and *data transfer blocks.* Each of these classifications is formed by a group of instructions of similar operation. Depending on the block type, there will be one or more control lines and one or more data specifications inside the block. An example of two generic blocks is shown in Fig. 7-11.

144

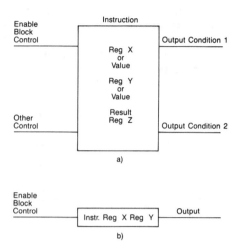

**Figure 7-11.** Example of generic blocks.

The following discussion and explanation of functional blocks in a generic manner will help the user to understand better the function and applications of these enhanced instructions. When dealing with functional blocks, the actual block described may differ slightly with any specific PC, but the function will be the same. It is almost impossible to present a block instruction that is common to all programmable controllers.

## Timer and Counter Operations

Some of the following timer and counter instructions are implemented in common ladder symbology; however, they are discussed here since some controllers have them in block form.

**Timers.** A typical timer block is shown in Fig. 7-12. This block may have one or two inputs depending on the PC. These input lines are normally labeled *control* and *enable*. If the control line is TRUE (i.e. has continuity), and the enable line is also TRUE, the block function will start timing.

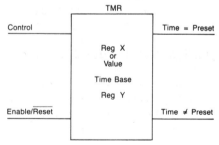

**Figure 7-12.**

Register X contains the preset time. The time base may be selectable in some controllers (0.01 sec, 0.1 sec, 1.0 sec, etc.), while fixed in others. Register Y stores the accumulated time during the timing. The timer block instruction will have one or two outputs; if there is only one output, it will be energized when the accumulated value equals the preset value. If there are two outputs, the other will be ON when the preset is not equal to the accumulated time (timer enabled but not timed-out). The output of the timer can contain a coil that is user-assignable.

If the timer has two input lines, it will time as long as the enable and control lines are TRUE. If the enable goes FALSE, the accumulated value will reset to zero. In some PCs, when the timer is timing and the control line is de-energized while the enable is one, the timing will stop and will resume timing at the accumulated value when the control line is energized again. These timers are known as *"interruptable"* timers. Timers that have only the control input will time when the control line is TRUE and reset the accumulated value when it goes FALSE. The timers with two inputs can act in this manner, if the control and enable lines are tied together with a branch (OR).

**Counters.** Counters are used in block form to count events or occurences of an input data when a transition from OFF-to-ON occurs. As in ladder diagrams, there are three types of counters: UP, DOWN, and UP/DOWN. A counter block is illustrated in Fig. 7-13.

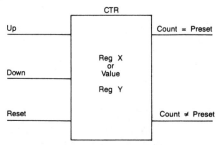

**Figure 7-13.**

The reset line is used to clear the accumulated count to zero. It can be used to reset the counter, based on any preceeding ladder logic representing the reset conditions. Register X is used to hold the preset count, and register Y stores the accumulated count. When the up input goes from OFF-to-ON, register Y is incremented and decremented at the OFF-to-ON transition of the down counter. There is no count while the reset line is one. There are two possible outputs that the counter controls: the first output is set to 1 when the preset count equals the accumulated count; the other is set when the count accumulation and preset are not equal.

## Arithmetic Operations

Most arithmetic block operations use three registers to operate. The first value is held in register X; another value in register Y; and the result of the operation is held in a third register, register Z. The numerical values will vary depending on the PC and are usually three, four, or five digits (BCD or binary). Typically, there is only one control input and anywhere from one to three outputs.

**Addition.** The addition functional block sums two values stored within the controller and places the results in a specified register. The values added can be fixed constants, values contained in I/O or holding registers, or variable numbers stored in any memory location. Figure 7-14 illustrates a typical addition functional block.

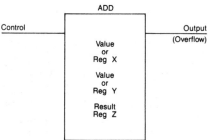

**Figure 7-14.**

The control line is used to enable the operation of the block. When the rung conditions are TRUE, the addition function will be performed. In this block, register X and register Y can be a preset value or a storage or I/O register. Each time the control signal is enabled (OFF-to-ON transition), the two numbers are added, and the result is placed in register Z. The output of the block is energized when the addition operation overflows. If the operation overflows, some controllers will clamp the results at the maximum value that the register can hold, while others will store the difference between the maximum count value and the actual overflow value.

Another method used by some controllers is the double precision addition with the same block. This operation is identical to the simple addition, but two registers are used to hold the numbers to be added, and two registers are used to store the result.

**Subtraction.** The subtraction functional block takes the difference of two values and stores the results in a result register. A typical subtraction functional block is shown in Fig. 7-15.

The control input operates in the same manner as in addition. When set to 1, the block operation is performed. The three operand registers are used to hold the data during the operation. The values these registers can assume also vary in format and may or may not include a sign. For example, register X could contain -9009 (decimal), and register Y could hold -10010. The result of this operation would be +1001, stored in register Z. The formats for subtraction vary, and sometimes the result register may not include a result sign. In this the case, the controller will normally provide three outputs: a positive result, register X greater than register Y; equal result, register X equal to register Y; or negative result, register X less than register Y. This type of block essentially performs a comparison function.

Some controllers use the subtraction block to read a numerical input value (e.g. analog, BCD, etc.) through I/O registers and to compare it to a set parameter. The output coils are used to signal a greater than, equal to, or less than the set point comparison.

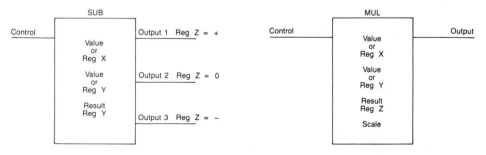

Figure 7-15.          Figure 7-16.

**Multiplication.** The multiplication block is illustrated in Fig. 7-16. The control line is used to enable the operation to take place. As with the previous math blocks, the multiplication functions have two registers to hold the operands and a register to store the result.

Normally, the product of two 4-digit numbers will result in an eight digit number. In such a case, some controllers will provide two registers in which to store the result. Other controllers use what is known as *scaling,* in which the result of the multiplication is held temporarily in two registers and then multiplied by the scale value. For example: if a PC has a 4-digit BCD format, and register X and Y contain 9001 and 8172 respectively, with a scaling value of -5 ($10^{-5}$), the result 73556172 will be held in two result registers temporarily (7355 and 6172), and then multiplied by the scaling value ($10^{-5}$). The result will be 735.56172. The result register will have 736 (rounded-off). The programmer should know that the result has been scaled. The actual result then is 736 X $10^5$ (73600000).

A block output is provided as an overflow signal, indicating that the result is greater than the result register can hold.

**Division.** The divide functional block performs the quotient calculation of two numbers that can be stored in one or more registers. This functional block is illustrated in Fig. 7-17.

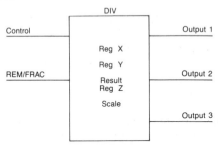

**Figure 7-17.**

The division begins with register X, the dividend, being divided by register Y, the divisor, and the result placed in register Z. The control line input will enable the block to perform the function. The REM/FRAC (REMainder/FRACtion) control line is used to determine whether one of the resultant register contents should be integer or fraction.

Some controllers also offer scaling factors to be specified in the block. This permits fractional results to be scaled and stored in a register. The fractional result would have been lost otherwise.

Depending on the PC used, there are three possible outputs that can be encountered. When energized, the top output generally represents a succesful division, the middle output will represent an overflow or error (divide by zero), and the lower output will indicate whether the result has a remainder or not.

**Square Root.** The square root functional block operates with one data register that varies in length depending on the block instruction. The square root operation is performed when the block is enabled.

## Data Manipulation Operations

Data manipulation instructions deal with operations that take place within one or two registers. Typical functions encountered in *data manipulation* include *logic matrix, data conversion, data comparisons, set constant parameters, increment registers, logical shift and rotate* operations, and *examine bit* functions.

**Logic Matrix.** The logic matrix functional block is used to perform AND, OR, EXCLUSIVE-OR, NAND, NOR, and NOT logic operations on two or more registers (see Chapter 3 for logic functions).

The performance of a logic function between two registers can be thought of as a matrix operation of length one, since there is one register for each operand. Figure 7-18 shows a typical block for the logic matrix function.

Defined by the block to be executed, the logic function is performed when the control input is enabled. The registers inside the block are user specified and generally are holding or storage registers. The result register Z will hold the result of the operation. The length indicates the number of words or registers adjacent to each of the operands X, Y, and Z, where data is located in matrix form.

As an example, let us assume a length equal to 10 (0 through 9) and the logical function to be AND. When the block is enabled, the contents of register X to

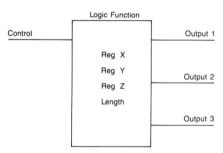

**Figure 7-18.**

register X + 9 will be ANDed with the contents of register Y to Y + 9; the result of the operation is placed in registers Z to Z + 9. Each register typically holds 16 bits of data; therefore, in this case, 160 bits in ten registers were ANDed with another 160 bits from registers Y to Y + 9, and the result (160 bits) stored in another matrix (registers Z to Z + 9).

There are three possible outputs in this block. The top output is energized once the control line is enabled, the middle output is energized when the operation is done, and the lower output is energized when there is an error.

Some controllers have only two operand registers (X and Y). When the logic operation is performed, the result is stored back in register Y, thus erasing the data previously stored in register Y. When this type of block is used, a data transfer to another register(s), prior to executing the logic matrix block, will prevent loss of the data.

**Data Conversions.** These instructions are provided so that the contents of a given register can be changed from one format to another and certain operations can be performed. Typical conversions include *BCD to binary, binary to BCD, absolute, complement,* and *inversion.*

BCD to binary code conversions are required when receiving BCD input data from field devices such as thumbwheel switches (TWS). A conversion in this case will allow the input data to be used in math operations (if required). The binary to BCD function operates in the same manner and is utilized when outputting data to field devices that operate in BCD, such as seven segment LED indicators. A typical functional block for this type of conversion is illustrated in Fig. 7-19.

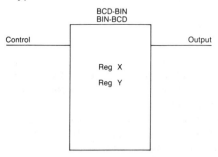

**Figure 7-19.**

The operation of this block, whether BCD-BIN or BIN-BCD, is basically the same. When the control input is enabled, the contents of register X (BCD or BIN) will be converted to binary or BCD depending on the conversion instruction. The results of the conversion are placed in register Y. The block output is energized when the instruction is completed.

The absolute, complement, and invert operations are generally done in a single register. In other words, the result of the operation will be stored back in the same location. A typical block is illustrated in Fig. 7-20.

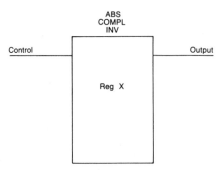

**Figure 7-20.**

The absolute block function computes the absolute value (always positive) of the contents of register X. If register X contains the value -5713, the result of X after the block instruction will be +5713.

The complement instruction, when executed, changes the sign of the contents of register X — positive to negative or negative to positive. For example, -6600 after execution will be +6600, and +3314 will be -3314.

The invert functional block inverts all the bits in register X. If the binary pattern in register X is 0000111100001111, after execution it will be 1111000011110000. The block output will be TRUE when the instruction is finished.

**Data Comparisons.** The compare functional block is used to compare the contents of two registers. A typical compare functional block is shown in Fig. 7-21. Registers X and Y contain the two values that will be compared. The output will be energized if the comparison has been satisfied (TRUE).

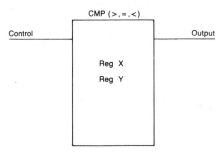

**Figure 7-21.**

Typical comparison instructions are greater than ($>$), less than ($<$), equal to ($=$), and combinations such as less or equal to, greater or equal to, and not equal to. Some controllers offer another comparison option, which uses another register (Z) to perform a limit (LIM) function. This limit instruction compares the value of register Y for less or equal to register X, and register Y greater or equal to register Z ($Y>X>Z$). If the compare instruction is TRUE, the output will be energized.

There are controllers that do not have a compare block, and yet perform a similar comparison using a subtraction block (see arithmetic blocks). In this case, there are three outputs that signal whether the result of the subtraction is positive (greater than), equal (equal to), or negative (less than).

**Set Constant Parameters.** It is sometimes necessary to store a constant parameter in a register that will be used later in the program for comparisons or set points. For this reason, some PCs provide a block instruction that allows a fixed value to be assigned to a register. A functional block for set constants is illustrated in Fig. 7-22.

When the block is enabled, register X is set equal to the value specified (in BCD,

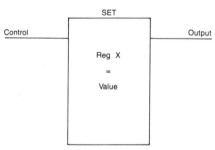

**Figure 7-22.**

binary, etc.), and the output is turned ON once the operation is completed. This instruction is very useful when resetting to zero during initialization of several storage or I/O registers.

**Logical Shifts and Rotates.** The shift instruction is used to move bits of a register(s) to the right or to the left. Figure 7-23 illustrates the execution of a right shift instruction. The left shift is identical, except the bit is moved in the opposite direction (shifting out the most significant bit).

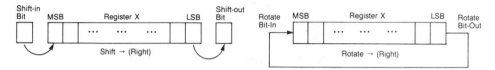

**Figure 7-23.** Illustration of a right shift operation.     **Figure 7-24.** Illustration of a right rotate operation.

The rotate instruction, like the shift instruction, shifts data to the right or left, but instead of losing the shift-out bit, it becomes the shift-in bit at the other end (rotate the bit). Figure 7-24 illustrates the operation of the right rotate instruction.

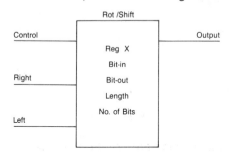

**Figure 7-25.**

A functional block for the shift or rotate function is illustrated in Fig. 7-25.

The control input enables the block operation for a rotate or shift execution. Some block instructions will have a right and left line to determine the shift or rotate direction. There are several variables that can be available inside the block, depending on the PC model. Register X is generally the location where the data to be shifted or rotated will be stored. If the length is to be specified, register X will be the starting location. For example, if the length is 10, then the block operation will be performed on 160 bits.

The bit-in and bit-out variables are used to specify the location of the bit whose value will be shifted in or out (available in shift block only). These bits can be real I/O locations that can be used to input or output data with the shift register operation. The number of bits indicate the amount of bit shifts or bit rotates that take place when the control input goes from OFF to ON.

151

**Examine Bit.** This functional block is used to examine the status of a single point, or bit, in a memory location. This type of instruction is generally used when *flags* are set during the program and then later tested and compared. A bit can be examined for an ON or OFF status. Figure 7-26 illustrates a typical block for this instruction.

The function is performed when the control input is TRUE. The bit position specified in register or location X will be examined for an ON or OFF condition. The output is energized if the instruction is TRUE.

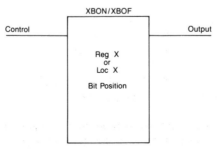

XBON/XBOF

Control                    Output

Reg X
or
Loc X

Bit Position

**Figure 7-26.**

## Data Transfer Operations

Data tranfers involve moving numerical data within the PC, either in single units or in blocks. Typical data tranfer blocks that are encountered are: *Move* (points and registers), *Move Blocks, Table Move, Block Transfers, ASCII Transfers,* and *FIFO Stack Transfers.*

**Move.** The move instruction is used to tranfer information from one location to another. The destination location will be to a single bit or register. A move functional block is shown in Fig. 7-27.

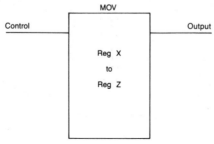

MOV

Control                     Output

Reg X

to

Reg Z

**Figure 7-27.**

When the control input is TRUE, the block operation is performed, and the contents of register X (or bit X) are copied into register Z (or bit Z). Register X, the *source* register, and register Z, the *destination* register, can be either a storage register or an I/O register. The output is energized when the instruction is completed.

In some PCs, the move function is performed on special word table locations. In this case, the data copied is automatically converted to the proper numerical format of the destination location. For example, register X, or word X, might contain a BCD value that when transferred to register Z, or word Z, is stored as a binary value, thus executing essentially a BCD to binary conversion with the move instruction.

Another type of move instruction involves the masking of certain bits within the register. This block is shown in Fig. 7-28. The data in register X is transferred to

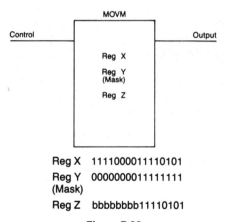

| Reg X | 1111000011110101 |
|---|---|
| Reg Y (Mask) | 0000000011111111 |
| Reg Z | bbbbbbbb11110101 |

**Figure 7-28.**

register Z, with the exception of the bits specified by a zero in the mask register Y. The bits *"b"* in register Z are set to zero due to the mask.

There is yet another move instruction that is found in several controllers. This is the *move status* instruction. With this block function, system status or I/O module status can be transferred to a storage register (register Z). The status information in the result register can then be masked, compared, or examined to determine the status of major or minor faults in the system or an I/O module and then used to take action under program control if necessary.

**Move Block.** The move block instruction causes a group of register or word locations to be copied from one place to another. The length of the block is generally user specified. Figure 7-29 illustrates a move block instruction.

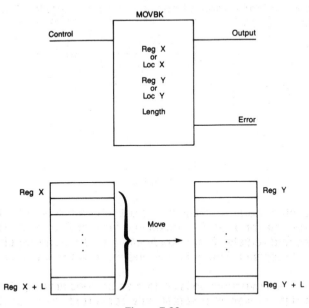

**Figure 7-29.**

When energized, the control input causes the execution of the block. The data starting at location X and through location $X+L$ (length $=L$) is transferred to locations Y through $Y+L$. The data in X to $X+L$ is left unchanged. Some PCs have the option of specifying how many locations can be transferred during one scan *(rate per scan)*.

153

**Table Move.** Table move instructions are associated with the tranfer of data from a block or table to a register or word in memory. There are basically two types of table moves: *table-to-register* and *register-to-table*. The main characteristic of this functional block is the manipulation of a pointer register, so that a register or word value can be stored in a particular table location as specified by the pointer. The table move is shown in Fig. 7-30.

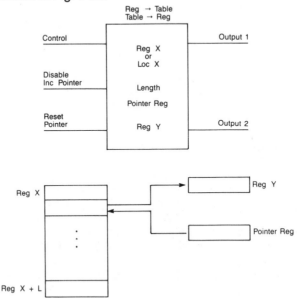

**Figure 7-30.**

When the control input goes from OFF-to-ON, the table move instruction is executed and the contents of the pointer register incremented. The middle input is used to disable the pointer increment. The bottom input of the table move block is used to reset the pointer to zero (initialize to top of table).

If it is necessary to store or retrieve data to or from a specific table location, the pointer register can be loaded with the appropriate value to point at the specified location. This loading is done prior to the table move through a set parameter or move register instruction.

The length L specifies the number of word locations in the table, counting from the starting location (X). The top output is energized when the block instruction is completed. The middle output of the table move block is energized when the pointer register has reached the end of the table.

Useful applications of the register to table transfer include the table loading of new data, storage of input information (e.g. analog) from special modules, or error information from the process being controlled. The table to register instruction is useful when changing preset parameters in timers and counters, and when driving a group of 16 outputs at one time through I/O registers.

**Block Transfer (In/Out).** The block transfer instruction is available in some PCs and primarily designed to be used with special I/O modules such as analog, encoder, BCD, and stepper motor. There are two basic types of block transfers that are used: *input* and *output.* Figure 7-31 shows a block transfer instruction.

The control input, when enabled, executes the block transfer instruction. The data on the I/O module (for a BKXFER IN) is stored in memory locations or registers starting at location X. The block length specifies how many locations are needed to store the I/O module data. For example, the data from an analog input module with four input channels can be read all at once, if the length is specified as four. The

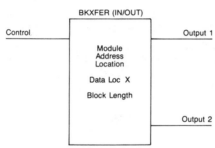

**Figure 7-31.**

block transfer output operates in a similar manner: the destination of the data transfer is determined by the address of the output module.

The top output of the block transfer instruction, when energized, signals the completion of the transfer operation. The bottom output is used to signal an error, such as bad I/O transmission or faulty module.

**ASCII Transfers.** The ASCII transfer instruction deals with the data transmission of ASCII characters from the PC to a peripheral device. This functional block operates in conjunction with an ASCII communications module, and its execution is under program control. The communication usually occurs two ways: *reading data* from a peripheral or *writing data* to a peripheral. This functional block is widely used when report generation is required in the application. Figure 7-32 illustrates a typical read/write ASCII functional block.

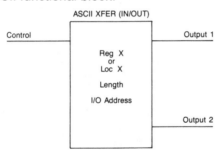

**Figure 7-32.**

The control input activates the ASCII transfer (In or Out). When reading data, the instruction will allow the special I/O module to perform a read function. The data is read from the module by the processor and then stored in special memory locations (X to length specified). The location of the module is indicated by the I/O address in the block. When writing data, the information is sent from the locations in which the message is stored to the address where the module is located.

Some ASCII transfer instructions have the capability of using a pointer register for accessing specific characters in the table (e.g. to decode a specific input character from the data table). Other ASCII instructions may allow the programmer to specify how many bytes or characters are transmitted during a scan. The speed of transmission (baud rate) is generally a function of scan time which can also depend on the number of ASCII devices active at one time. The ASCII transfer instruction assumes that proper baud rates, start/stop bits, and parity have been set-up in the I/O module.

**FIFO Stack Transfers.** The *First-In-First-Out* (FIFO) instruction is used to construct a table or queue where data is stored. The basic function of this operation is similar to an asynchronous shift register, in which one word (16 bits) is shifted within the stack each time the instruction is executed. The data is shifted in the same order in which it is received — the first one in will be the first one out. The FIFO operation is analogous to the idea of first come (In) first serve (Out).

The FIFO operation generally consists of two instruction blocks: *FIFO IN* and *FIFO OUT.* FIFO IN is used to load the queue and FIFO OUT unloads the queue. The FIFO moves are useful for storing, and later retrieving, large groups of temporary data as it comes in. A typical application of the FIFO instruction is to store and retrieve data that is synchronized to the external movement of pieces or parts on a conveyor or transfer machine. Figure 7-33 shows a typical FIFO block instruction.

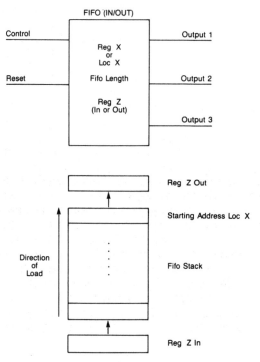

**Figure 7-33.**

The FIFO block is initiated by the OFF-to-ON transition of the control input logic. The reset signal may be available for resetting the FIFO stack *(clear stack).* For the FIFO IN instruction, register Z holds the data which will be transferred to the queue. This data is placed in the FIFO stack (bottom location) when the control input is TRUE. When using the FIFO OUT block, the data is output through register Z. The length of the stack is specified by FIFO length (register X to X+length).

There are generally three outputs used in the FIFO blocks. The top output indicates that the functional block is operating; the middle output signals that the queue is empty; and the bottom output shows the queue is full. Most controllers have both the input and output FIFO blocks; however, some PCs may offer both instructions (IN and OUT) in one FIFO functional block.

## Other Functional Blocks

Other blocks found in some programmable controllers do not fall into any of the previous categories; such blocks are *Sequencers, Diagnostics,* and *PID control.*

**Sequencers.** The sequencer block is a powerful instruction that is used to simulate a drum timer. A sequencer is analogous to a music box mechanism in which each peg produces a tone as the cylinder rotates and strikes the resonators. In a sequencer, each peg (bit) can be interpreted as a logic 1, and no-peg as a logic 0. Sequencers are generally specified, using a table which is similar to the music box cylinder spread-out. Depending on the controller, the number of bits can vary

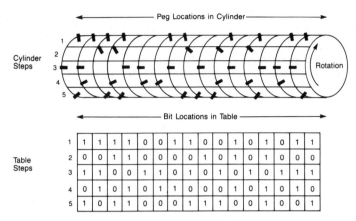

**Figure 7-34.** Music box cylinder and sequencer data table comparison.

from 8 to 64 or more. Figure 7-34 illustrates a cylinder and sequencer table comparison.

The width of the table may vary, as does the size of the music cylinder. Each of the steps in the sequencer can be an output (through I/O registers) representing each of the pegs. Figure 7-35 shows a typical functional sequencer block.

The block is initiated by the OFF-to-ON transition of the control input, and the contents of the sequencer table will be output in a sequential manner. Each step being output is pointed to by the contents of the pointer register. Every time the control input is energized, the pointer register is automatically incremented and points to the next table location. The reset pointer input is provided in case the pointer register needs to be reset to zero (point to step 1). The sequence length and width specify respectively how many steps and bits are used in the table. The output of the block is energized whenever the sequencer instruction is enabled.

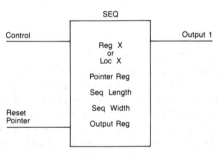

**Figure 7-35.**

**Diagnostics.** This block instruction performs the comparison of two memory blocks in which one contains actual input conditions and the other contains *reference* conditions. This comparison is done on a bit-by-bit basis to find if the blocks are identical. If a miscomparison occurs, the bit number and the state of the bit are stored in a holding register. This block is illustrated in Fig. 7-36.

The diagnostic block is performed when the control input is energized. The contents of locations X through length L are compared with the contents of the reference locations Y through length L. If a difference is found, it is stored in register Z. The contents of location Y are not altered. The top output is energized when the instruction is complete, and the second output is on when a miscomparison is found.

Predetermined behavior of inputs and outputs (reference conditions) is generally determined by the machine being controlled. However, there are some control-

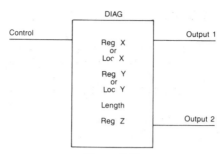

**Figure 7-36.**

lers that allow the reference conditions to be taught to the PC. These controllers input the reference *"teaching"* conditions using sequencer input intructions, block transfers in, and other instructions depending on the model used. Diagnostic instructions are useful for signaling the operator of a machine (not PC) malfunction, such as limit switch or other components.

**PID.** The PID functional block is used in PCs that offer the capability of performing analog control using the Proportional-Integral-Derivative (PID) algorithm. The user specifies certain parameters associated with the algorithm to control the process correctly. Figure 7-37 illustrates a typical PID block.

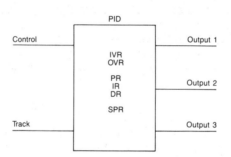

**Figure 7-37.**

The control input is generally used to enable the PID block to operate (auto). The bottom input *track*, when energized, will determine if the PID variables are being tracked, but not output. If the block is not enabled (in manual mode), the controller can still track the variables when track is enabled. The *input variable register (IVR)* and *output variable register (OVR)* are user specified and are associated with the location of the analog modules (input and output). The *proportional register (PR), integral register (IR),* and *derivative register (DR)* hold the gain values that need to be specified for the control of the process. The *set point register (SPR)* holds the target value for the process set point. Depending on the controller, other block variables can be specified, such as dead times, high and low limits, and rate of update. The top output of the PID block is used to indicate an active loop control, while the middle and bottom outputs usually indicate a low and high limit alarm respectively.

Some controllers provide PID capabilities without having the block instruction. Generally, in this case, a special PID module is used in which all the input/output parameters are in the module. The set point and gain parameters are transferred to the module during initialization of the program through an output instruction, such as block transfer out or move data to an output register. The module data can be altered under program control if any of the parameters need to be changed.

158

## 7-6 ENGLISH STATEMENT LANGUAGE

English statement languages for programmable controllers can be considered a derivative of computer languages. Compared with languages that many computers use, the PC English statement instructions are much easier and more operator oriented. These English statements, or *control statements* as they may also be called, have provided more computing power to the controller that was previously using only ladder or Boolean instructions.

The high-level control statements provide an advantage in the manipulation of data and numerical I/O, since data can be inserted in registers directly through simple English instructions. Peripheral communication for reporting can also be simplified when using these languages, since an instruction such as PRINT can be used to output a string of data.

These control statements have been adopted in larger PCs whose manufacturers did not take the functional block approach. Advocates of control statements as a high-level language give two main reasons for their support: the statements' simplicity eases the programming of a control task, and their English-type instructions allows other users to easliy interpret the program once it has been written. There are generally a fixed number of instructions available, and they differ from one manufacturer to another; nevertheless, in one way or another, they accomplish the same task.

There are several types of control statements that are presently used in the industry. Most of these languages are similar to the widely used BASIC language used in personal computers. Some manufacturers, such as GTE-Sylvania (model MC-16) use a language called SYBIL (see Fig. 7-38), which is very similar to the BASIC language. However, the number of instructions used by the MC-16 are reduced. The SYBIL instruction set is summarized in Table 7-3.

```
 1 REM   SC(6,2)=1 ==> START OF MAIN PROGRAM
 2 REM   SC ARE INPUT MODULES, AI IS ANALOG INPUT
 3 REM   LD ARE OUTPUT MODULES
 5 DECLARE A,B=WORD
10 LET A=300
20 IF SC(6,2)=0 GO TO 10
30 IF SC(6,1)=1 IF SC(4,2)=0 IF SC(4,3) GO TO 80
40 IF SC(6,1)=0 B=AI(2,1)
50 B=B/2
60 IF B=A LET LD(3,1)=1
70 GO TO 100
80 WAIT SEC=10
90 GO TO 30
100 REM CONTINUATION OF PROGRAM
```

**Figure 7-38.** Example of SYBIL control program.

Cincinnati Milacron's Mechanism Control Language, MCL (model APC-2), can be used to program a complete control program without having any basic level language (i.e. ladder diagrams or Boolean). MCL has English-style words that can be used to identify variables (input/output) in plain English. These user specified words operate with the MCL instruction set to provide a powerful method of programming. One characteristic of this system is that the programming unit (CRT) compiles the control statement program before going into the controller unit,

**Table 7-3.** Summary, SYBIL Instruction Set (GTE Sylvania)

**DECLARE (variable)** = WORD

This instruction is used during initialization to store the variables named. The declared variables store a number ranging from -32767 to +32767.

**DIM A(X)**

This statement reserves an area in memory for the number of elements specified in the array size X.

**FOR (variable)** = X **to** Y   **STEP** Z

This instruction allows the setting of initial and final conditions of a variable. The instruction is executed as many times as specified from X (value) to Y (value) in increments (STEP) of Z (value). It performs a LOOP function where a routine is done several times.

**NEXT (variable)**

The next intruction is used with the FOR statement at the end of the LOOP routine. It indicates an increment of X by a step Z until X equals Y, when the LOOP is finished.

**GO SUB** nnn

The GO SUBROUTINE instruction is used to call a routine in the program where control is transferred. The subroutine is identified by the line nnn of the program.

**RETURN**

The RETURN statement is used at the end of the subroutine called by GO SUB nnn. This instruction returns program control to the instruction that follows the GO SUB nnn in the main program.

**GO TO** nnn

The GO TO statement causes an unconditional jump to statement nnn in the program.

**IF** (condition) **THEN** (another statement)

This instruction allows conditional execution of another statement.   If the condition is TRUE, the other instruction is executed. If FALSE, the program jumps to the next program line.

**INPUT** (variable)

The INPUT statement is used to read the value(s) from the CRT terminal and store the value in the variable defined.

**LET (variable = expression)**

This instruction causes the value of the variable to become equal to the value of the expression. It can also be used for arithmetic functions and control equation solutions. The LET statement is optional, and may be omitted.

**PRINT** (expression)

This statement prints the value of an expression or variable.

**PRINT** "(expression)"

This PRINT instruction prints the expression inside the quotation marks and can be used for report generation.

**REM** (comment)

The REMARK statement is used for documentation purposes and is not executed by the program. It will print any comment in the control program printout for clarification.

**WAIT** (Unit) = (Amount)

The WAIT statement provides a delay execution of the program.   The units can be 20 msec, seconds, minutes, hours, or days.

thus speeding execution of the program. Other BASIC languages may use an interpreter in the system which makes program execution slower.

Another manufacturer that offers an English statement-like language is Reliance Electric. This instruction set, called Control Statements, is similar to BASIC, but is control oriented and *syntax free*. Syntax free means that when the desired instruction is entered, all the syntax is displayed in the proper places, and blanks are left for any parameters that must be entered. Instructions are designed to allow easy handling of field devices. Another characteristic of their AutoMate controller is that the basic relay and timer/counter instructions are handled in the ladder language, while more complex control functions are performed by the control statement language. Each language is handled by a separate processor, with interaction between the two to perform the total control task. In general, other PCs using high level statements solve the ladder program in the control statement language. A partial list of this control statement instruction set is given in Table 7-4.

**Table 7-4.** Summary, Control Statement Instruction Set (Reliance Electric)

| **Math Statements** | | These statements perform the indicated arithmetic operation on two 16-bit numeric variables and place the result in a third variable. |
|---|---|---|
| ADD | Addition | bles and place the result in a third variable. |
| SUB | Subtract | The double precision DADD and DSUB statements allow calculations or more than 16 bits to |
| MUL | Multiply | ments allow calculations or more than 16 bits to |
| DIV | Divide | be handled.   All calculations are in fixed point, |
| DADD | Double Precision Add | with provisions made for moving the decimal |
| DSUB | Double Precision Subtract | point. |

| **Input/Output Statements** | | |
|---|---|---|
| DIN | Digital Input | The input statements read data from the appropriate input module and place the data into the |
| AIN | Analog Input | priate input module and place the data into the |
| FIN | Frequency Input | designated location. The output statements |
| DOUT | Digital Output | take data from a designated location and output |
| AOUT | Analog Output | the data to the appropriate output module. |
| FOUT | Frequency Output | |

| **ASCII Communications Statements** | | |
|---|---|---|
| PRINT | Print | The first three statements allow text, values of |
| PRINTV | Print Value | variables, or the contents of strings to be output |
| PRINTS | Print String | via a serial peripheral communications inter- |
| READC | Read Character | face. The last two statements allow individual |
| READS | Read String | characters or complete strings to be input via a serial peripheral communications interface. |

| **Logical Statements** | | |
|---|---|---|
| AND R | AND Register | These statements perform the indicated logical |
| OR R | OR Register | operation on two 16-bit registers and place the |
| XOR R | Exclusive OR Register | results in a third register. Similar instructions allow operations on bit or byte variables. |

| **Compare Statements** | | |
|---|---|---|
| CMP = | Compare Equal To | These statements perform the indicated com- |
| CMP > | Compare Greater Than | parison between two variables and set a flag bit |
| CMP < | Compare Less Than | if the comparison is true. |

| **Single Register Operations** | | |
|---|---|---|
| SHR | Shift Right | These statements perform the indicated opera- |
| SHL | Shift Left | tion on a single 16-bit register. |
| ROR | Rotate Right | |
| ROL | Rotate Left | |
| INC | Increment | |
| DEC | Decrement | |

**Table 7-4.** Summary, Control Statement Instruction Set (continued)

### Put Statements

| | | |
|---|---|---|
| **PUT B** | Put Binary | These statements load a specified constant into |
| **PUT D** | Put Decimal | a variable in the indicated format. |
| **PUT A** | Put ASCII | |
| **PUT P** | Put Point | |

### Move Statements

| | | |
|---|---|---|
| **MOVR** | Move Register No Conversion | These statements move data from one 16-bit |
| **MOVR B** | Move Register BCD-Binary | register/byte to another, performing the indi- |
| **MOVR D** | Move Register Binary-BCD | cated conversion. |
| **MOVR A** | Move Register Absolute Value | |
| **MOVR C** | Move Register Complement | |
| **MOVR I** | Move Register Invert | |
| **MOVR S** | Move Register (value) To String (ASCII) | |
| **MOVS R** | Move String (ASCII) To Register (value) | |
| **MOVE B** | Move Byte | |

| | | |
|---|---|---|
| **MUX** | Multiplex | These statement allow data to be multiplexed |
| **DMUX** | Demultiplex | from many registers to one or demultiplexed from one register to many. |

### Array Handling Statements

| | | |
|---|---|---|
| **STRING** | Define String | These statements reserve space in the applica- |
| **BINTBL** | Define Binary Table | tion memory for array variables. These arrays |
| **DECTBL** | Define Decimal Table | can contain 16-bit numerical data (Tables) or ASCII characters (STRINGS). |

| | | |
|---|---|---|
| **GETSTR** | Get String | These statements allow data to be read from or |
| **PUTSTR** | Put String | written in individual locations within the speci- |
| **GETTBL** | Get Table | fied table or string. |
| **PUTTBL** | Put Table | |

| | | |
|---|---|---|
| **DATA** | Data | This statement allow a table or string to be initial-ized with data. All the locations in the array can be specified. |

### Flow Of Control Statements

| | | |
|---|---|---|
| **GOSUB** | Go To Subroutine | These statements allow the normal sequence of |
| **GO TO** | Go To | statement execution to be altered. Subrou- |
| **I/OINT** | I/O Interrupt | tines can be called, interrupts can be serviced, |
| **SQINT** | Sequence Interrupt | and sections of the program can be skipped |
| **RET** | Return | over. |
| **IF** | If | |

### Comment Statement

| | | |
|---|---|---|
| **REM** | Remark | This statement allow non-executing comments to be entered into the program for clarity. |

The operation of most of these high-level instructions is very similar in concept to the functional block instructions.

## 7-7   CHOOSING THE RIGHT LANGUAGE

As in most cases in which there is a choice, pros and cons are associated with each choice, and so it is with choosing a PC language. Each language will have its advantages as well as its limitations. In the previous sections, we discussed how ladder diagrams and Boolean, as well as functional blocks and English statements, could accomplish similar functions. Both ladders and Boolean are easy to use, and there are those who would argue a similar statement for both high-level languages. The best way then to select the language which meets a user's needs is to take into account certain factors, including the following:

- the ease of use and implementation
- the primary characteristics of the language
- the type of problem to be solved
- the execution-time requirements that will be imposed on the program

Before discussing these criteria, it should be understood that selecting a PC language is not quite like selecting a language for a computer. A computer language can be selected and purchased after already owning a computer, but the selection of a PC language is actually a part of selecting the controller itself. In other words, when a user buys the PC, it already has a language. "Choosing the right language," then, simply means evaluating the language along with the controller to see if it meets established requirements.

*The ease of use* of a particular language is a consideration that should take into account not only those users presently responsible for implementing the control system, but also personnel who will be expected to make necessary changes and to maintain the system. Presently, all PC languages are for the most part reasonably easy to use; however, the choice of a language will ultimately depend on who is going to use and maintain it.

Ladder diagram logic, the oldest and most conventional PC language, is formed from relay-contact logic symbols. For anyone with any knowledge of relay-logic, this language will be second nature and require little training. An advantage it offers is that when displayed on the CRT programmer or on a hardcopy printout, it takes the same form as a relay drawing. Trouble-shooting a ladder program is much like finding problems on a relay drawing, but much simpler.

In contrast to the users who would probably prefer ladder diagrams are those users who are totally unfamiliar with relay-logic, and yet have an understanding of Boolean logic and high-level languages. The preference in this case may be Boolean mnemonics, but its acceptance by others should be considered. A similar preference situation would probably arise with functional blocks and statement-oriented languages. Being of a ladder format, functional blocks might be more appealing and easier to understand for those users familiar with relay-logic. Whatever language is selected, the time and and labor cost associated with the development, installation, and maintenance of the program will be related to how well it is understood and accepted.

In deciding on the language, both *the primary characteristics* of the languages available and the type of problems to be solved should be evaluated. The pertinent questions are: *what exactly are the prime functions of each language,* and, *is the language designed to solve a particular problem or perform a specific task?* An answer to these questions will spell out which language is best suited to the problem. The ladder language, for example, was designed to solve logic networks involving simple timing, counting, and sequencing. A ladder diagram program to perform these functions is easily written and will be as efficient as any user-oriented language could possibly be. If complex decoding or math routines (using basic operators) or communicating with peripherals is required, the task may be impossible or may require brute methods to accomplish.

If the application calls for "number crunching," control of special devices, peripheral communication, complex data handling, or other such items, be sure the language is flexible enough to handle them efficiently.

Finally, the execution-*time requirements* of a program are considerations that emphasize the scan time of a program written in a particular language and the amount of memory that will be required to store it. If the selected language is properly suited to the application, only a minimum number of instructions will be required to perform the various tasks. Conversely, a language improperly suited to the application will require more instructions than would a better suited language. The latter case would be exemplified if a high-level language like BASIC is used to solve large amounts of relay-ladder logic. A ladder diagram program would require less memory and execute much faster.

The high-level languages become better suited to applications that involve a variety of control functions (e.g. logic, communication, numerical manipulations, control of data I/O, etc.). English-like statement languages are also advantageous in that programs tend to be self-documenting, but disadvantageous to the extent that programs can become wordy, lengthy, and require excessive amounts of time for execution and debugging.

Choosing the best language to solve a particular task is not difficult, but should not be taken lightly. If a good choice is made, the resultant program is likely to take less time to create, to require less storage, to execute at an acceptable speed, and to solve the control problem satisfactorily. If a poor choice is made, programming time, storage use, and execution time may be excessive. The program may never perform as intended and may fail to provide an acceptable solution to the control problem.

# CHAPTER
# --]8[--

# PROGRAMMING THE
# CONTROLLER

*A good system design does not insure a good
program, but a poor system design goes a long
way towards insuring a bad one.*

The task of developing the control program is not difficult, but it involves much more than simply writing instructions. Regardless of the size of the programming task, several development steps must be carried out before an effective solution to the problem can be achieved. In this chapter, we discuss these program development steps and put them to use in two programming examples. Finally, several programming tips are provided for some frequently encountered problem situations.

## 8-1 DEFINING THE CONTROL TASK

A user should begin the problem-solving process by defining the task and making sure of what needs to be done. To understand a task is to understand the scope of the task. It requires determining the type of input data that is to be used and the type of output required. This information, in turn, provides a basis for determining the programmable controller operations that must be performed.

The definition of the control task should be done by those who are familiar with the operation of the machine or process. This limitation will help minimize possible errors due to misunderstanding of the process application. Usually, task definition must occur at many levels. Individuals within each department involved must be consulted to determine what inputs are required or have to be provided, so that everybody understands what is to be done in the project. For example, in a project involving the automation of a manufacturing plant in which materials are to be retrieved from the warehouse to the automatic packaging area, personnel from both the warehouse and packaging areas must collaborate with the engineering group during the system definition. In case there are data reporting requirements, management should also be involved.

If a task is currently done manually or through relay logic, steps of this procedure should be reviewed to determine what improvements if any are possible. Although relay logic can be directly implemented in a PC, it is advisable, when possible, to redesign the procedure to meet current needs of the application and to take advantage of the capabilities that a programmable controller offers.

The factors to be considered when defining the task are all closely related to the success or failure of the resultant program. This relationship will be revealed in the ability to provide correct control of the machine or process.

## 8-2 DEFINING THE STRATEGY

After the task has been defined, the planning for its solution can begin. This procedure commonly involves determining the sequence of processing steps that must take place within a program to produce the output controls. This part of the program development is known as the development of an *algorithm*.

The term algorithm may be new or strange to some readers — it need not be. Each of us follows algorithms to accomplish certain tasks in our daily lives. The procedure that a person follows to get from home to school or work is an algorithm; he or she exits the house, gets into the car, starts the engine, and so on; in the last of a finite number of steps, the destination is reached.

The strategy implementation for the control task using a PC closely follows the development of the algorithm. The user must implement the control from a given set of basic instructions and produce the solution, or answer, in a finite number of such instructions. In most cases, it is possible to develop an algorithm to solve a problem. If doing so becomes difficult, it may be that further definition is needed. In this case, a return to the problem definition step may be required. For instance, we cannot specify how to get from where we are to Kokomo, Indiana, unless we know both where we are and where Kokomo is. If a particular method of transportation is required, we need to be told that information as part of the problem definition. If there is a time constraint, we need to be told also.

A fundamental rule in defining the program strategy is: *think first, program later.* Consider alternative approaches to solving the problem. Allow time to polish the approach (solution algorithm) selected before trying to program the control function. Adopting this philosophy will shorten the programming time, reduce debugging time, accelerate the start-up, and permit the focusing of attention on design when designing and on programming when programming.

During the formulation stage of the strategies, the user will be faced with a new application or a modernization of an existing process or machine. Regardless of which application must be done, the user will have to review the sequence of events that takes place, and through the addition or deletion of steps, optimize the control. Input and output considerations should be addressed, and a knowledge of what field devices the PC will control is required.

## 8-3   A SYSTEMATIC APPROACH

A programmable controller is a powerful machine that can do only what it is told to do, and do it only as it is told to do so. It receives all its directions from the control program, the set of instructions or solution algorithms created by the programmer.

A successful PC control program depends greatly on how organized the user is. There are many ways to approach a problem, but if the application is approached in a systematic manner, the probability of making mistakes lessens.

The techniques used for the implementation of the control program are subject to the individual involved in the programming. However, it is always recommended that certain guidelines be followed. Table 8-1 shows two approach guidelines that have proven to be very useful for implementing a programmable control system. One of the approaches is directed towards new systems, while the other is for modernizing (upgrading) the contol system for existing systems that are functioning without a PC (e.g. relay or card-logic control).

**Table 8-1.** Approach guidelines for New Applications and Modernizations

| New Applications | Modernizations |
|---|---|
| • Understand the desired functional description of the system | • Understand the actual process or machine function |
| • Review possible control method(s) and optimize the process operation | • Review machine logic of operation and optimize when possible |
| • Flowchart the process operation | • Assign real I/O addresses and internal addresses to inputs and outputs |
| • Implement the flowchart by using logic diagrams or relay logic symbology | • Translate relay ladder diagram to PC coding |
| • Assign real I/O addresses and internal addresses to inputs and outputs | |
| • Translate the logic implementation to PC coding | |

As mentioned previously, understanding the process or machine operation is the first step in a systematic approach to solving the control problem. For new applications, the planning of strategy will follow the problem definition. Reviewing strategies for new applications, as well as revising the actual method of control for a modernization, will help detect or minimize possible errors that were introduced during the planning stages.

The differences in approach between a new or a modernization project become apparent during the programming stage. In a modernization project, the user ordinarily understands thoroughly the operation of the machine or process and what needs to be controlled. The sequence of events is usually defined by an existing relay ladder diagram, like the one shown in Fig. 8-1, that can be translated into PC ladder diagrams almost directly. However, a procedure should be followed to avoid possible errors and to maintain organization.

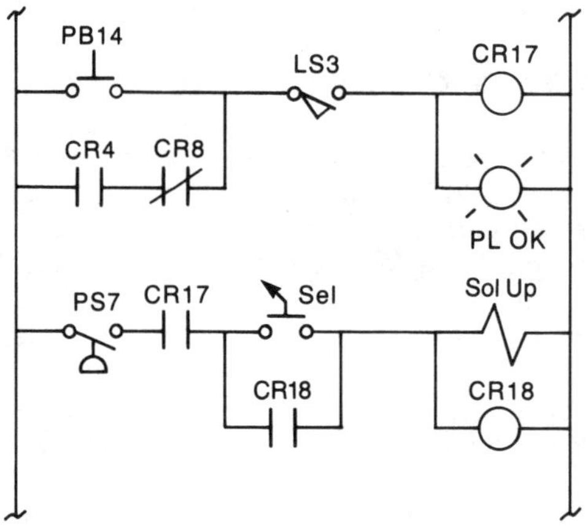

**Figure 8-1.** Partial relay ladder diagram of a typical existing system.

New applications usually begin with specifications given to the person or persons who will design and install the control system. These specifications are translated into a written description that explains the possible forms of control. The written explanation should be in simple terms so confusion is avoided.

*Flowcharting* is one technique often used in planning a program after a written description is made. A flowchart is a pictorial representation that serves as a means of recording, analyzing, and communicating problem information. Broad concepts, as well as minor details and their relationship to each other, are readily apparent. Sequences and relationships that are hard to extract from general descriptions become obvious when displayed on a flowchart. Even the flowchart symbols themselves have specific meanings, which aid in the interpretation of the solution algorithm. Figure 8-2 illustrates the most common flowchart symbols and their meaning.

Flowcharts usually describe the process of operation in a sequential manner. A simple flowchart is illustrated in Fig. 8-3. Each step in the chart performs an operation, whether an input/output, decision, or processing of data.

Once the flowchart is completed, the logic sequences can be obtained in one of two ways. First, *logic gates* specifying the input conditions, whether real or internal, can be used to describe a particular output sequence. Second, *PC contact symbology* can be used directly to implement the logic necessary to represent an output rung. Figure 8-4 illustrates these two methods. Users should employ

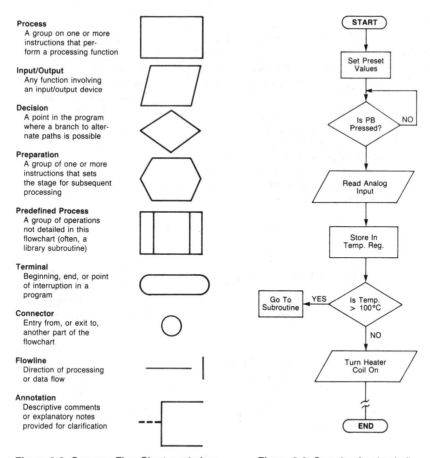

**Process**
A group on one or more instructions that perform a processing function

**Input/Output**
Any function involving an input/output device

**Decision**
A point in the program where a branch to alternate paths is possible

**Preparation**
A group of one or more instructions that sets the stage for subsequent processing

**Predefined Process**
A group of operations not detailed in this flowchart (often, a library subroutine)

**Terminal**
Beginning, end, or point of interruption in a program

**Connector**
Entry from, or exit to, another part of the flowchart

**Flowline**
Direction of processing or data flow

**Annotation**
Descriptive comments or explanatory notes provided for clarification

**Figure 8-2.** Common Flow Chart symbols.

**Figure 8-3.** Sample of a simple flow chart.

whichever method feels most comfortable. However, logic gate diagrams may be more appropriate if the controller uses a Boolean instruction set.

The inputs and outputs marked with an X in Fig. 8-4a, indicate real I/O in the system. If no marking is present, the I/O point is an internal. The designations for the actual input signals can be the actual devices (e.g. LS1, PB10, AUTO, etc.) or symbolic letters or numbers that are associated with each of the field elements. A short description of the sequence can be helpful later during programming and is strongly recommended during this stage.

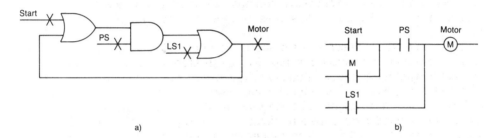

a)

b)

**Figure 8-4.** Logic implementations: **(a)** logic gate diagrams and **(b)** contact symbology diagram.

**Address Assignment.** The assignment of inputs and outputs is one of the most important procedures that takes place during the programming stages. The I/O assignment will document in an orderly fashion and organize what has been done thus far. It will indicate what PC input is connected to what input device and what PC output will drive what output device. The assignment of internals, including timers, counters, and MCRs, also takes place here. These assignments are the actual contact and coil representations that are used in the ladder diagram program.

The assignment of real inputs and outputs, as well as internals, can be tabulated as shown in Table 8-2. The numbers associated with the I/O address assignment depend on the PC model used. These addresses can be represented in octal, decimal, or hexadecimal so the count assignment will vary depending on the number system used by the PC.

The description part of these assignment tables is used to describe the input or output field devices (real I/O) as well as internal use (internals). The I/O assignment can be extracted from the logic gate diagrams, or ladder symbology, that were used to describe the logic sequences.

The table of assignments is closely related to the input/output connection diagram shown in Fig. 8-5. Although currently there are no industry standards for input and output representations, they are typically represented by squares and diamonds, respectively. The middle section is used for *coding* the PC program as will be shown later.

**Table 8-2a.** Sample I/O address assignment document for real inputs and outputs.

| Module Type | Rack | Group | Terminal | Description |
|---|---|---|---|---|
| | **I/O Address** | | | |
| Input | 0 | 0 | 0 | LS 1 — Position |
| | 0 | 0 | 1 | LS 2 — Detect |
| | 0 | 0 | 2 | Sel Switch — Select 1 |
| | 0 | 0 | 3 | PB 1 — Start |
| Output | 0 | 0 | 4 | PL 1 |
| | 0 | 0 | 5 | PL 2 |
| | 0 | 0 | 6 | Motor M1 |
| | 0 | 0 | 7 | Sol 1 |
| Output | 0 | 1 | 0 | Sol 2 |
| | 0 | 1 | 1 | PL 3 |

**Table 8-2b.** Sample address assignment document for internal outputs.

| Device | Internal | Description |
|---|---|---|
| CR7 | 1010 | CR7 Replacement |
| TDR10 | T200 | Timer on Delay 12 sec |
| CR10 | 1011 | CR10 Replacement |
| CR14 | 1012 | CR14 Replacement |
| — | 1013 | Set-up interlock |
| • | • | • |
| • | • | • |
| • | • | • |

A conscious grouping of associated inputs and outputs is recommended during the I/O assignment. This grouping will allow monitoring or manipulation of a group of I/O (through I/O registers) simultaneously. For instance, if 16 motors are to be started sequentially, their starting sequence can be viewed by monitoring the I/O register associated with the 16 I/O points (mapped). Due to the modularity of the I/O system, it is also recommended that all the inputs or all the outputs be assigned at the same time. This practice will prevent assignment of an input address to an output module and vice versa.

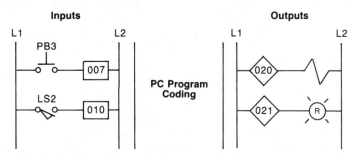

**Figure 8-5.** Typical input/output connection diagram.

**Portions To Leave Hardwired.** During the assignment of inputs and outputs, it is necessary to make the decision as to what elements should not be wired to the controller. These elements will remain part of the magnetic control logic and usually will include elements that are not frequently switched off after start, such as compressors, hydraulic pumps, and others. Elements like emergency stops (e.g. rope switches, push buttons, etc.) and master start push buttons should also be left hardwired, principally for safety purposes. If for some reason the controller is faulty and there is an emergency situation, the system can be shut down without PC intervention.

Figure 8-6 illustrates an example of components that are typically left hardwired. Note that in this diagram the PC fault contact (or watchdog timer contacts) is wired in series with other emergency conditions. These contacts are closed when the controller is operating correctly and would open when a fault occurs.

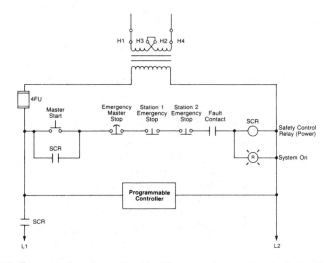

**Figure 8-6.** Example of sections of logic left hardwired and not controlled by the PC.

171

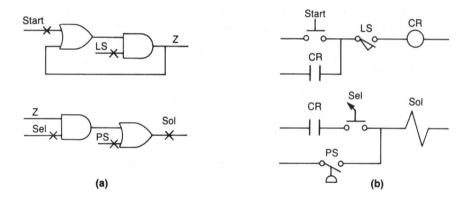

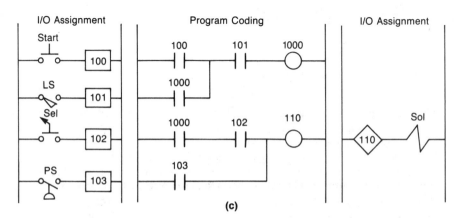

**(c)**

**Figure 8-7.** Illustration of program coding **(c)** generated from logic gates **(a)** and relay logic **(b)**.

**Coding.** Coding is the process of writing or rewriting the logic or relay diagram into PC ladder program form. This ladder program is the actual logic that will implement the control of the machine or process and is stored in the the application memory. The ease of program coding is directly related to how orderly the previous stages (assignment, etc.) have been done. Each element in the PC ladder program has an address assigned to it according to the I/O assignment document. Figure 8-7 shows a sample program that is generated from the logic or relay diagram (internal coil 1000 replaces CR).

Note that the coding is a PC representation of the logic, whether it comes from a new application or a modernization. This coding process is examined closer in the next two sections, where two programming examples are presented: one a modernization and the other a new application.

**Programming An Input Condition.** During the program coding stages, the user will find that the programming of a normally closed (NC) or normally open (NO) input device (e.g. limit switch, push button, etc.) is not necessarily programmed as the same state. Figure 8-8 illustrates a normally closed push button wired to a pilot light in hardwired logic and its equivalent PC I/O wiring and coding.

During the operation of a programmable controller, the processor scans the inputs and tests for different conditions. If the input has been programmed as normally open, the PC tests or examines for an ON or CLOSED condition to energize the output. If the input has been programmed normally closed, the PC tests or examines for an OFF or OPEN condition to energize the output.

172

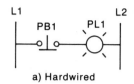

a) Hardwired

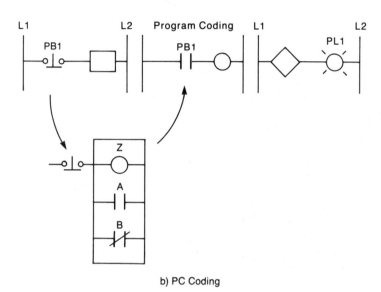

b) PC Coding

**Figure 8-8.** An illustration of PC program coding **(b),** to accomplish the same function as the hardwired logic in **(a).**

In the hardwired logic, the pilot light (PL1) is ON as long as PB1 is not pushed (see Fig. 8-8a). The same operation must be performed by the control program (Fig. 8-8b). The input PB1 is wired to the input module as a normally closed push button, but the program coding is normally open. Since the pushbutton is normally activated (allowing power to the input module), the processor will see a closed (1) condition when the instruction is evaluated. A "1" will close the normally open contact, and the pilot light will illuminate until the pushbutton is depressed.

To visualize this operation better, think of the input module as having a coil Z which has a contact A, normally open, and contact B, normally closed. These contacts are used by the PC to allow PB1 to be evaluated as either NO or NC. If coil Z is energized, contact A will be closed, and B will be opened. If PB1 is normally closed, the normal state of coil Z will be ON, or energized. When PB1 is pushed, coil Z will de-energize. As can be seen, PB1 will keep coil Z ON, and the state of contacts A and B during normal operation will be closed and open, respectively. This ability to examine a single device for either an open or closed state is the key to the flexibility of a PC; however the device is wired (NO or NC), the controller can be programmed to perform the desired action without changing the wiring.

In most cases, an input device wired normally closed is programmed as a normally open PC contact. Remember that the programming state of an input depends not only on how it is wired, but also on the desired control action. Figure 8-9 shows an example in which one push button, with two contacts, is brought to the PC and is programmed differently depending on which contact is wired to the module.

173

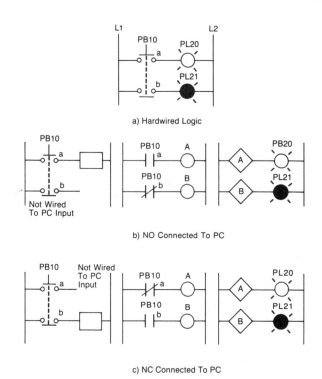

**Figure 8-9.** Illustration of program coding of a dual contact input, depending on the contact that is wired to an input module.

In this example, PB10 has two contacts, NO and NC. The NO contact is used to energize PL20, while the NC contact energizes PL21 (Fig. 8-9a ). When implementing this circuit in a PC, only one input needs to be brought to the controller, since the signal from PB10 can be programmed as NO or NC. Programming of the contacts depends on the desired logic and which contact is connected to the input module. If the NO contact of PB10 is wired to the input module (PB10A), the coding is done as shown in Fig. 8-9b. If the NC contact of PB10 is wired to the module (PB10B), the program coding is reversed as shown in Fig. 8-9c. The program logic is inverted   because during normal operation (no pressing of PB10) the PC contact PB10A will open, and PB10B will close.

## 8-4  PROGRAMMING EXAMPLE 1: A MODERNIZATION

Modernization applications often involve the transformation of a machine or process control from conventional relay logic to a programmable controller. The relay panels that contain the control logic are usually characterized by maintenance problems involving contact chatter, contact welding, and other electromechanical problems. Changes, targeted to improve performance in the machine can also call for optimization of the control with a programmable controller.

This example presents a modernization in which the machine control is to be changed from hardwired relay logic to PC programmed logic. The field devices to be used are the same, with the exception of those that can be implemented in the controller (e.g. timers, control relays, interlocks, etc.). The objectives of modernizing the control of this machine are: a *more reliable control system, less energy consumption, less space used* by the control panel, and a *flexible system* that can

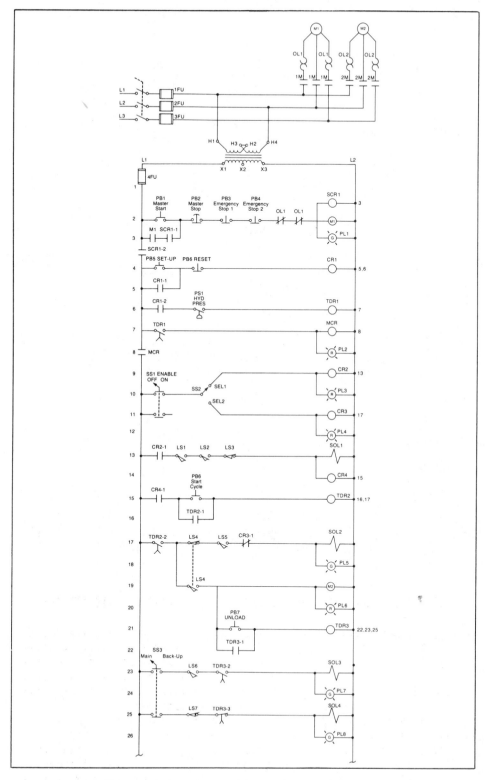

**Figure 8-10.** Relay ladder diagram for modernization example.

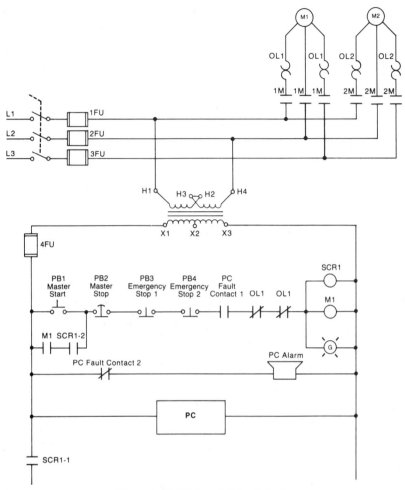

**Figure 8-11.** Portions left hardwired.

accommodate future expansion. Figure 8-10 illustrates the relay ladder diagram that presently controls the logic sequencing for this particular machine. For the sake of simplicity, only part of the total relay ladder logic is shown.

Initial review of the relay ladder diagram indicates that certain portions of the logic should be left hardwired (lines 1, 2, and 3). This practice will keep all emergency stop conditions independent of the controller. The hydraulic pump motor (M1), which is energized only when the master start is pushed (PB1), is also left hardwired.

The safety control relay (SCR) will provide power to the rest of the system if M1 is operating properly, and no emergency push button is depressed. Furthermore, we can include the PC fault contact in series with the emergency push buttons and also connect it to a PC failure alarm. The portions to leave hardwired are shown in Fig. 8-11. Note that during proper operation, the PC will energize the fault coil (PCFC), thus closing PCFC1 and opening PCFC2.

Continuing the example, we can now start assigning the real inputs and outputs to the I/O assignment document. All control relays, as well as timers and interlocks from control relays, will be assigned internal output addresses. Table 8-3 shows the assignment and description of each input and output. The address assignment defines the termination point of each field device connected to the controller. Each address is defined by the *rack number,* the *group number,* and the *termination point.*

176

**Table 8-3(a).** Real I/O address assignment document for modernization example.

| Module Type | I/O Address | | | Description |
| --- | --- | --- | --- | --- |
| | Rack | Group | Terminal | |
| Input | 0 | 0 | 0 | PB 5 — Set up PB |
| | 0 | 0 | 1 | PB 6 — Reset (Wired NC) |
| | 0 | 0 | 2 | PS 1 — Hydraulic Pressure Switch |
| | 0 | 0 | 3 | SS 1 — Enable Selector Switch (NC contact left unconnected) |
| Input | 0 | 0 | 4 | Sel 1 — Select 1 position |
| | 0 | 0 | 5 | Sel 2 — Select 2 position |
| | 0 | 0 | 6 | LS 1 — Limit Switch Up-Position 1 |
| | 0 | 0 | 7 | LS 2 — Limit Switch Up-Position 2 |
| Input | 0 | 1 | 0 | LS 3 — Location Set |
| | 0 | 1 | 1 | PB 6 — Start Load Cycle |
| | 0 | 1 | 2 | LS 4 — Trap (Wired NC) |
| | 0 | 1 | 3 | LS 5 — Position Switch |
| Input | 0 | 1 | 4 | PB 7 — Unload PB |
| | 0 | 1 | 5 | SS 3 — Main/Back-up (Wired to PC NO) |
| | 0 | 1 | 6 | LS 6 — Max. Length Detect |
| | 0 | 1 | 7 | LS 7 — Min. Length Back-up |
| | 0 | 2 | 0 | |
| | 0 | 2 | 1 | |
| | 0 | 2 | 2 | Spare |
| | 0 | 2 | 3 | |
| | 0 | 2 | 4 | |
| | 0 | 2 | 5 | |
| | 0 | 2 | 6 | Spare |
| | 0 | 2 | 7 | |
| Output | 0 | 3 | 0 | PL 2 — Set-up OK |
| | 0 | 3 | 1 | PL 3 — Select-1 |
| | 0 | 3 | 2 | PL 4 — Select-2 |
| | 0 | 3 | 3 | Sol 1 — Advance FWD |
| Output | 0 | 3 | 4 | Sol 2 — Engage |
| | 0 | 3 | 5 | PL 5 — Engage On |
| | 0 | 3 | 6 | M 2 — Run Motor |
| | 0 | 3 | 7 | PL 6 — Motor Run On |
| Output | 0 | 4 | 0 | Sol 3 — Fast Stop |
| | 0 | 4 | 1 | PL 7 — Fast Stop On |
| | 0 | 4 | 2 | Sol 4 — Unload with Back-up |
| | 0 | 4 | 3 | PL 8 — Back-up On |

The number system used is octal, and the maximum I/O capacity is 512 I/O (000-777 octal). Note that the inputs with multiple contacts such as LS4 and SS3 are only connected once to the PC. The I/O interface modularity is 4 points (circuits) per module; there can be 2 modules per group and 8 groups per rack (total of 64 I/O per rack). This PC configuration is shown in Fig. 8-12.

The program coding and I/O assignment for this example is shown in Fig. 8-13. This ladder program illustrates several special coding techniques that must be used to implement the PC logic. Among those techniques used are the MCR function in software, instantaneous contacts from timers, OFF-delay timers, and the separation of rungs having multiple outputs.

**Table 8-3(b).** Internal output address assignment document for modernization example.

| Device | Internal | Description |
|--------|----------|-------------|
| CR1 | 1000 | CR-1 |
| TDR1 | T100 | Timer preset 10 sec Register 3000 ACC Register 3001 |
| MCR | MCR 200 | First MCR Address |
| CR2 | — | Same as PL3 Address |
| CR3 | — | Same as PL4 Address |
| CR4 | — | Same as Sol 1 |
| — | 1001 | To set-up internal for instantaneous contact of TDR2 |
| TDR 2 | T101 | Timer preset 5 sec Register 4002 ACC Register 4003 |
| — | 1002 | To set-up internal for instantaneous contact of TDR3 |
| TDR 3 | T102 | Timer preset 12 sec Register 4004 ACC Register 4005 |

Internal addresses start at 1000
Timers addresses start at T100 - to - T137
MCR addresses start at MCR 200 - to - MCR 217

The MCR internal output is used to accomplish a function similar to the hardwired MCR. Referring to the relay logic diagram, if the MCR is energized, its contacts will close and allow power to be provided to the rest of the system. In PC software, this same function is accomplished by using the internal MCR200 (for this example, MCR200 is the first available address for MCRs). If the MCR coil is not energized, the PC will not execute the ladder logic that is "fenced" between the MCR coil and the END MCR instruction.

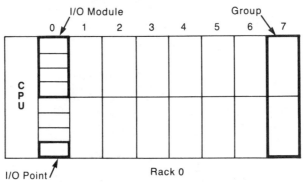

**Figure 8-12.** Block illustration of PC for modernization project.

The control relay CR2, in line 9, does not need to be replaced with an internal, since the "contacts" of PL3 (line 10) can be used. This technique can be used whenever a control relay is in parallel with a real output device.

The separation of the coils in lines 17 and 18 of the hardwired logic is done, since the PC used here does not allow rungs with multiple outputs. Having separate rungs for each output is also a good practice.

The normally-closed inputs that are connected to the input modules are programmed normally-open, as explained in the previous sections. The limit switch LS4 has two contacts (NO and NC in lines 17 and 19 of Fig. 8-10). However, only one set of contacts needs to be connected to the controller. In this example, we have selected the normally-closed LS4. Although the normally-open contact is not connected to the controller, its hardwired function can still be achieved by programming LS4 as a normally-closed ladder contact (refer to programming an input device in Section 8-3).

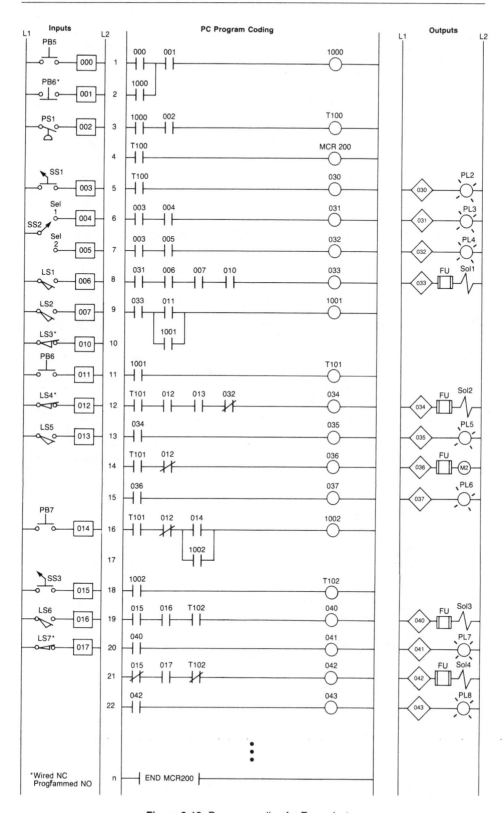

**Figure 8-13.** Program coding for Example 1.

179

Applications such as this one also require timers that have instantaneous contacts, which are not available in most PCs. An instantaneous contact is one which opens or closes when the timer is enabled. To overcome not having an instantaneous contact, an internal coil can be used. Line 15 in the hardwired logic shows that if PB6 is pressed and CR4 is closed, the timer TDR2 should start timing, and contact TDR2-1 would seal PB6. This arrangement cannot be implemented in the PC without special considerations. If we use the software timer contacts, the timer will not seal until it has timed-out. If PB6 is released, the timer will be reset because PB6 is not sealed. A solution to the problem is accomplished by using internal coil 1001 to seal PB6 and start timing T101 (TDR2). This technique is shown in the PC program coding in lines 9, 10, and 11. The time delay contacts of T101 are used for the ON or OFF delays.

## 8-5 PROGRAMMING EXAMPLE 2: A NEW APPLICATION

New applications generally start with a written description of the control task. This application example deals with the automation of a batching process. It will include the process description, controller requirements, flowchart, the logic diagrams for each output sequence, assignment of I/O, and the program coding.

**Description.** Figure 8-14 illustrates the process flow diagram that describes the elements that will be controlled in this batching application. There are two ingredients (A and B) that will be mixed in the reactor tank. The reactor tank must be empty as indicated by the normally closed liquid level switch (LLS) and at a temperature of 100°C before ingredient A can be added. The mixer motor should be off to avoid liquid precipitation, and the finished product tank should be in position as detected by a limit switch.

Once the initial temperature is set, ingredient A will be added to the tank by opening solenoid valve 1 (SOL1), until 100 gallons have been poured in. Detection of gallon quantity is achieved by LLS1, which is normally open. This switch closes when the proper level is reached. At this point, ingredient B is added by opening SOL2. The quantity required of ingredient B is 400 gallons, and it is detected by LLS2 (500-100 = 400 gal.). The temperature should be kept at 100°C, ±10%, during the add ingredients step. If the temperature drops, the heater should be turned on automatically, while the process continues.

After both ingredients have been added to the reactor tank, the temperature should be raised to 300°C (±10%). Then the mixer should be turned ON for 20 minutes. The temperature should be controlled automatically at set point during the process.

The drain valve should be activated by SOL3 after the mixing is completed. This operation should reset the process until another tank is placed in position, and the cycle is started again.

Pilot light indicators should be incorporated in the system as a means of alerting the operator of the status of the batching process.

**Controller Requirements.** This application requires a controller that has the capability of reading analog signals from the process. In this case, the voltage comes from a temperature transducer (0-10 volts), which has a range of 0°C to 500°C (50°C/volt). The temperature is controlled by the ON/OFF control of the heater coil. Standard 110 VAC input and output modules are required.

**Process Flowchart.** A process flowchart is shown in Fig. 8-15. It illustrates in global terms what has been described in the definition of the control task and serves as a preparation for the logic diagrams.

**Logic Diagrams.** The initial implementation of the logic required to control each of the process sequences can be accomplished with the use of logic diagrams. These diagrams represent the necessary conditions that will enable a rung to be

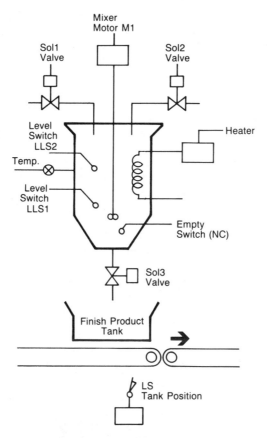

**Figure 8-14.** Process flow diagram for Example 2.

energized. The real I/O is marked with an X. Figure 8-16 shows the logic diagrams for this example.

The first logic diagram represents the initial requirements to start the process. The start push button, when pressed, enables the start mix output, if the tank is in position, SOL3 is closed, and the stop push button is not pressed. The pilot lights PL1 and PL2 indicate that the tank is in position, and the system START MIX signal is enabled.

Logic diagram 2 starts the setting of the initial temperature (T1) at 100°C. The logic indicates that the mixer motor (M1) must be off, START MIX enabled, and the reactor tank empty. The READY TO MIX input is an interlock from logic diagram 6, and it is used in logic diagram 2 to disable T1 when T2 is being set. Note, the EMPTY signal is used in the OR function with the initial SET T1 to insure that even when the tank is still adding ingredients, the temperature control will continue to mantain the temperature at T1.

Logic diagram 3 controls the READY TO ADD signal, which allows ingredient A to be added. Here, the output TEMP OK1 (T1 = 100°C) of the block indicates that the temperature has been reached, and that START MIX signal is still enabled; thus, the process is ready to add the first ingredient.

In logic diagram 4, the READY TO ADD A enables SOL1 to open. This action occurs while LLS1 is still opened (less than 100 gallons), and the drain valve (SOL3) is not energized. When the liquid level is reached, LLS1 closes and according to the logic, it de-energizes SOL1. The second part in the logic diagram indicates that the add ingredient A step is finished.

In logic diagram 5, the SOL2 is opened to add ingredient B until LLS2 is closed (500 gallon level) indicating that 400 additional gallons have been added to the reactor tank. The remainder of the logic indicates that the add ingredient B step is finished.

Logic diagram 6 shows that once both ingredients are in the reactor tank (finish A, and finish B, both ON), the READY TO MIX control signal is enabled. This condition will start a new temperature control block to raise the temperature to 300°C and will disable the other temperature control (T1).

In logic diagram 7, after the temperature is at 300°C, and the READY TO MIX (SET T2) is on, the mixer will turn on. At the same time, the timer is enabled, and after 20 minutes (1200 sec), it times-out and resets the mixer motor logic. The timer output represents the FINISH MIXING signal, used to energize SOL3, which opens the drain valve to discharge the mixed ingredients (logic diagram 8). The valves remains open until the empty switch is back to its normal state (closed).

The logic diagram 9 is used to turn the heater on, if the temperature is low. The heater can be turned on from either of the two temperature control block outputs. The sequences 10, 11, and 12 are provided to indicate to the operator the status of the temperature inside the tank.

The logic for reading the temperature will be done using *compare* functional block instructions. Once the command, or logic, indicates control of temperature, the compare block will be enabled, and three comparisons will be done to detect more than 110°C, 100°C, and less than 90°C.

It will be necessary to compare using a *limit (LIM)* compare function, since it is required that the ingredients be added at 100°C. The output of this block will be OK1. The logic for the pilot light indicates to the operator that the temperature is OK. This logic is the combination of the NOT at 110°C or greater, AND NOT at 90°C or less; thus the range is within the tolerances as specified (100°C ± 10%). This logic is shown in Fig. 8-17. The limit instruction also applies to the control of T2 (temperature), with the exception of the setpoint (see coding).

**Assignment of I/O.** The assignment of real I/O can begin by addressing the real inputs and outputs. These inputs and outputs are tabulated in the assignment table. The assignment of I/O for this application example is illustrated in Table 8-4. Note that the modularity for the digital I/O is four points per module. The analog module contains two input channels that occupy one half of a group (four locations).

The assignment of internals is shown in Table 8-5. Here, several internal coil addresses are tabulated to represent control relay conditions related to the logic diagram. The coils associated with the compare functional blocks are internals used to describe the temperature conditions, such as HI, and LO. The analog value of the temperature is stored in I/O register 3034 and will be compared to storage registers that hold the equivalent values of the temperature ranges.

**Coding.** Coding is the next step after the tabulation of the input and output assignments. The coding of the program is done by following the logic diagram sequences previously specified and also by using the I/O and internal tables as reference for the addresses. The program coding for this example is shown in Fig. 8-18.

Note that the ladder logic shown in the program coding is the implementation of each logic diagram. The internals are assigned as specified in the internal assignment table. Several storage registers are added in the compare blocks to hold the preset values. These values correspond to the equivalent temperature set points used, including the tolerances (i.e. 110°C, 100°C, 300°C, 270°C, etc.).

The voltage from the temperature transmitter is 0 to 10V, representing 0 to 500°C. Each volt represents a change of 50°C. The controller used in this example receives the voltage and converts it to a count ranging from 0000 to 9999. This is proportional to the voltage and, therefore, to the temperature.

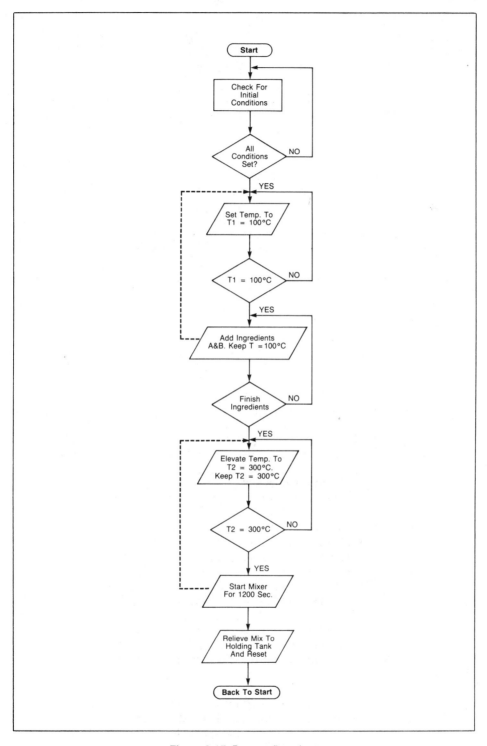

**Figure 8-15.** Process flow chart.

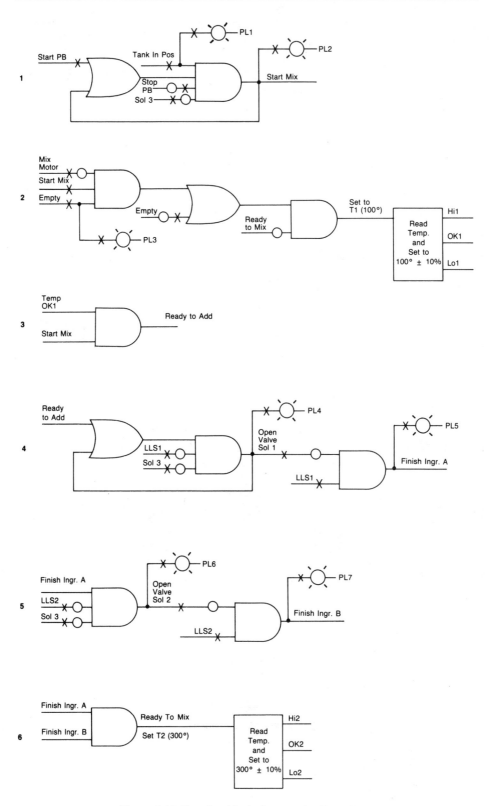

**Figure 8-16.** Functional logic diagrams for Example 2.

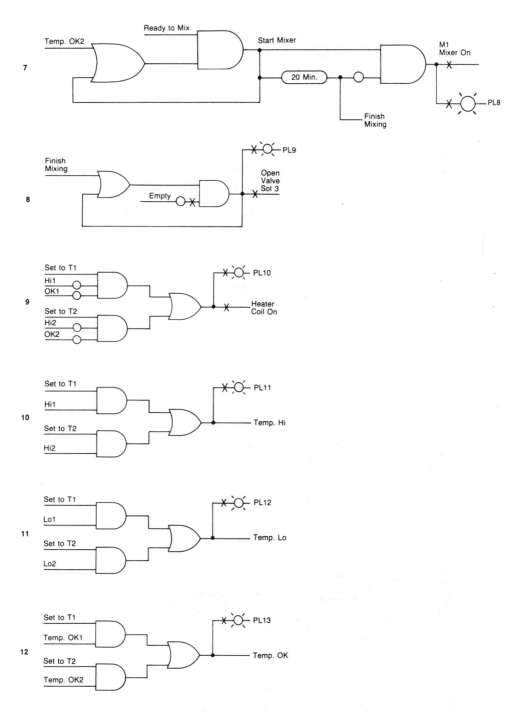

**Figure 8-16.** Continued.

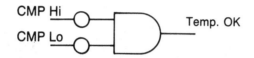

**Figure 8-17.** Logic diagram for TEMP OK signal.

**Table 8-4.** Address assignment document for real I/O in Example 2.

| Module Type | I/O Address Rack | Group | Terminal | Description |
|---|---|---|---|---|
| Input | 0 | 0 | 0 | Start Mix PB |
| | 0 | 0 | 1 | Stop PB |
| | 0 | 0 | 2 | Tank Position LS |
| | 0 | 0 | 3 | Empty Switch (NC) |
| Input | 0 | 0 | 4 | LLS 1 (100 gal) |
| | 0 | 0 | 5 | LLS 2 (500 gal) |
| | 0 | 0 | 6 | Not Used |
| | 0 | 0 | 7 | Not Used |
| Output | 0 | 1 | 0 | PL Tank in Position |
| | 0 | 1 | 1 | Start Mix |
| | 0 | 1 | 2 | Empty Reactor Tank |
| | 0 | 1 | 3 | Solenoid Valve 1 (Ingredient A) |
| Output | 0 | 1 | 4 | PL Valve 1 Open |
| | 0 | 1 | 5 | Finish A |
| | 0 | 1 | 6 | Sol Valve 2 |
| | 0 | 1 | 7 | PL Valve 2 |
| Output | 0 | 2 | 0 | Finish B |
| | 0 | 2 | 1 | Mixer (M1) |
| | 0 | 2 | 2 | PL Mixer On |
| | 0 | 2 | 3 | Solenoid Valve 3 (Drain) |
| Output | 0 | 2 | 4 | PL Valve 3 Open |
| | 0 | 2 | 5 | Heater Coil |
| | 0 | 2 | 6 | PL Heater On |
| | 0 | 2 | 7 | PL Temp H |
| Output | 0 | 3 | 0 | PL Temp Lo |
| | 0 | 3 | 1 | PL Temp OK |
| | 0 | 3 | 2 | Not Used |
| | 0 | 3 | 3 | Not Used |
| Analog Input | 0 | 3 | 4 | Input 34 Connected to Transmitter |
| | 0 | 3 | 5 | Corresponds to I/O Register 3034 |
| | 0 | 3 | 6 | Input 36 left as is |
| | 0 | 3 | 7 | Register 3036 |

The first set point (register 4000) contains the value 2200 which indicates a count proportional to 110°C (100°C + 10%); register 4003 contains 1800, equivalent to 90°C (100°C -10%). Registers 4001 and 4002 contain values of 2040 (102°C) and 1960 (98°C) respectively. These values are used to detect a small range area in which the temperature is very close to 100°C to start the ingredient addition. We do not compare the value to 100°C because this value may never be exactly 100°C at a particular time while reading the analog value. These two registers are used in a LIM compare block that detects when the temperature is between 98°C and 102°C. The preset values of the other compare blocks are specified in the same manner.

**Table 8-5.** Address assignment document for internal outputs in Example 2.

| Device | Internal | Description |
|--------|----------|-------------|
| Logic | 1000 | Set to T1 |
| CMP Block | 1001 | CMP Hi 1 |
| LIM | 1002 | CMP for range OK1 |
| CMP Block | 1003 | CMP Lo 1 |
| Logic | 1004 | TEMP OK 1 |
| Logic | 1005 | READY TO ADD |
| Logic | 1006 | READY TO MIX/SET to T2 |
| CMP Block | 1007 | CMP Hi 2 |
| LIM | 1010 | CMP for range OK2 |
| CMP Block | 1011 | CMP Lo 2 |
| Logic | 1012 | TEMP OK2 |
| Logic | 1013 | START MIXER |
| Timer | T100 | Timer preset 20 min (1200 sec) Register 4000 ACC Register 4001 |

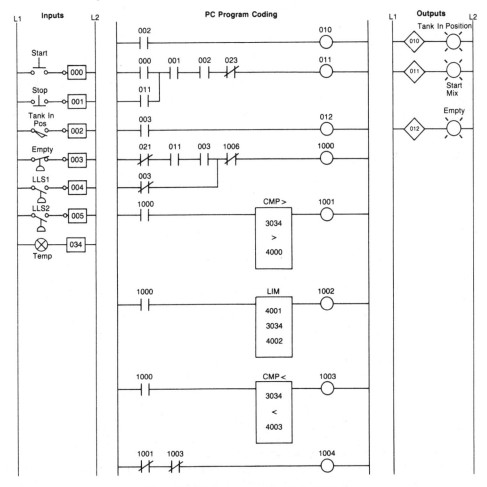

**Figure 8-18.** Program coding for Example 2.

**Figure 8-18.** Continued.

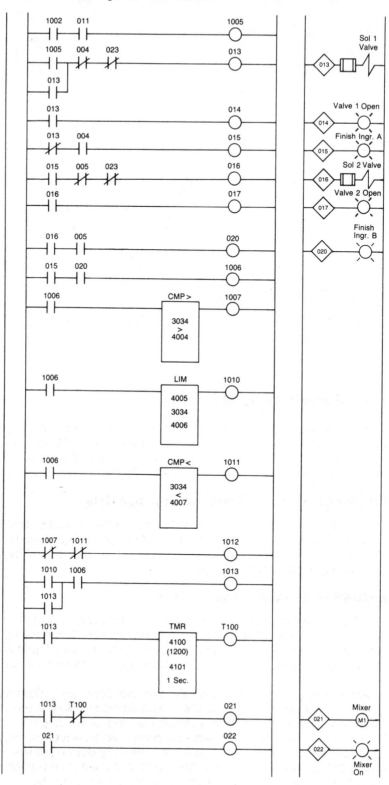

**Figure 8-18.** Continued.

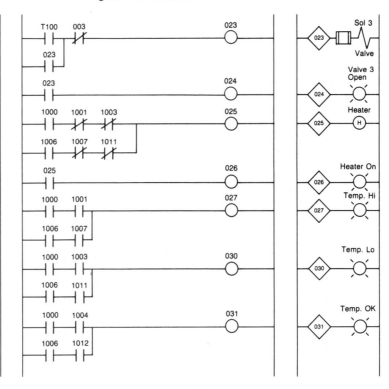

## 8-6   PROGRAMMING TIPS

This section will present several examples of logic networks that are often required when programming the controller. For convenience, the examples are implemented using the most basic ladder diagram instructions, and therefore may require more instructions than would some higher level instruction sets.

### EXAMPLE 1: Using Internal Storage Bits

Most programming devices are limited in the number of series contacts, or parallel branches, that a rung can have. This limitation can be overcome by using internal storage bits, as shown in Fig. 8-19. This same technique would have been applied if the contacts had been in series.

### EXAMPLE 2: Start/Stop Circuit

The Start/Stop circuit shown in Fig. 8-20 can be used to start or stop a motor or process or to simply enable or disable some function. To start a motor, the ladder output need only reference the motor output address. If the intent is only to detect that some process is enabled, the output can be referenced with an internal address.

Note in Fig. 8-20 that the Stop PB and the Emergency Stop inputs are programmed normally-open. They are programmed this way, since these types of inputs are usually wired normally-closed. As long as the Stop PB, and the E. Stop PB, are not pushed, the programmed contacts will allow logic continuity. Since the Start pushbutton (normally-open) is a momentary device (allows continuity only when closed), a contact from the motor output is used to seal-in the circuit. Often, the seal-in contact is an input from the motor starter contacts.

189

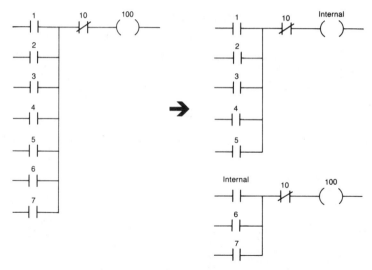

**Figure 8-19.** Using internal storage bits.

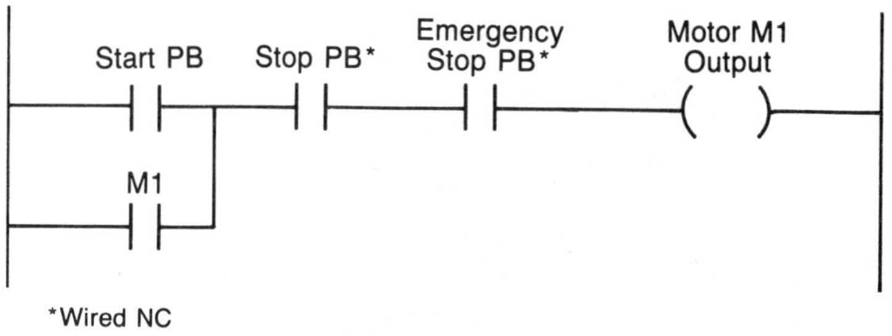

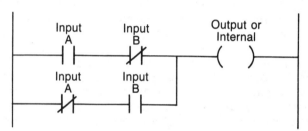

**Figure 8-20.** Start/Stop circuit.

## EXAMPLE 3: Exclusive-OR Circuit

The Exclusive-OR circuit in Fig. 8-21 is used when it is necessary to prevent an output from energizing if two conditions, which can activate the output independently, occur simultaneously.

**Figure 8-21.** Exclusive OR circuit.

## EXAMPLE 4: The One-Shot Signal

The one-shot (transitional output) in Fig. 8-22 is a program generated pulse output that, when triggered, goes HIGH for the duration of one program scan and then goes LOW. The one-shot can be triggered from a momentary signal such as a pushbutton, or from an output that comes on and stays on for some time (e.g. motor). Whichever signal is used, the one-shot is triggered by the leading edge (OFF-to-ON) transition of the input signal. It stays ON for one scan and goes OFF. It stays OFF, until the trigger goes OFF, and comes ON again. A one-shot is typically used as a clear or reset signal. The one-shot signal is perfect for this application, since it stays ON for only one scan.

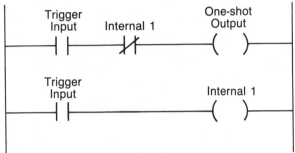

**Figure 8-22.** One-Shot pulse circuit.

## EXAMPLE 5: Initialization Using An MCR

The logic circuit shown in Fig. 8-23 can be used when it is necessary to set up several parameters during an initialization period. Typically, this initialization is done only once during the program, either when the system is first powered up, or when power is reapplied after a power loss. The parameters that are usually initialized are timer and counter preset values, high and low limit setpoint values, or any other preset or starting values.

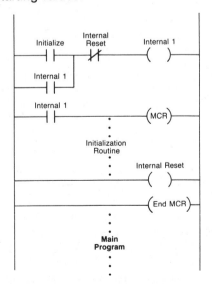

**Figure 8-23.** Initialization circuit using an MCR.

191

## EXAMPLE 6: System Start-UP Horn

The start horn logic (Fig. 8-24) is often used when moving equipment is about to be started (e.g. conveyor motors). The SET-UP signal in this example is similar to the start/stop circuit, but instead of starting the system, it enables the timer, which allows the horn to sound for 10 seconds. Note that the horn sounds when the START input is closed, and until the timer times out or the RESET input opens. The system can be started if the SET-UP signal remains ON, and the horn delay timer times out.

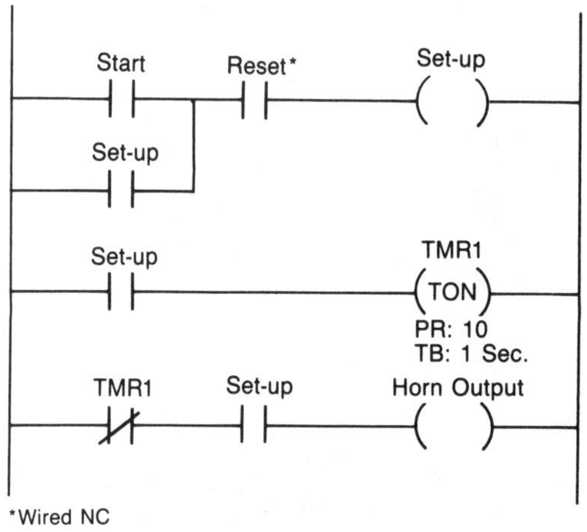

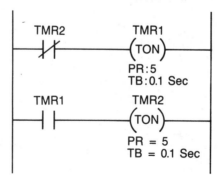

*Wired NC

**Figure 8-24.** Circuit for system start-up horn.

## EXAMPLE 7: Oscillator Circuit

The oscillator logic (Fig. 8-25) is a simple timing circuit that can be used to generate a periodic output pulse of any duration. This pulse is generated by TMR1.

**Figure 8-25.** Oscillator (ON/OFF) circuit.

## EXAMPLE 8: Annunciator Flasher Circuit

The flasher circuit, shown in Fig. 8-26, is used to toggle an output ON and OFF continually. The "oscillator circuit" (from example 7) output (TRM1) is programmed in series with the alarm condition. As long as the alarm condition is TRUE, the annunciator output will flash. The output in this case would be a pilot light; however, this same logic could be used for a horn that is pulsed during the alarm condition. Note that any number of alarm conditions could be programmed using the same flasher circuit.

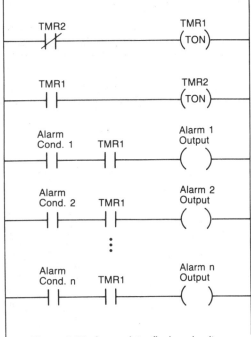

**Figure 8-26.** Annunciator flasher circuit.

## EXAMPLE 9: Self-Resetting Timer

The self-resetting timer (Fig. 8-27) will provide a one scan pulse each time the timer is energized. The repetition of this pulse is determined by the specified preset value of the timer.

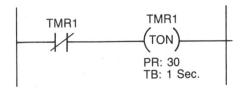

**Figure 8-27.** Self-resetting timer circuit.

## EXAMPLE 10: Sequential Motor Start

This example (Fig. 8-28) illustrates how several motors or other devices could be started sequentially, as opposed to all at once. For simplicity in this example, we use an ON-delay timer to delay the start of each motor. However, this approach would be impractical for starting a large number of motors. If a large number of motors are to be started, other techniques that do not require as many timers as motors are used (e.g. shift registers, self-resetting timers, oscillator circuits, etc.).

## EXAMPLE 11: Delayed De-Energize Device

This example (Fig. 8-29) illustrates the use of the OFF-delay timer to de-energize a motor or any device after a delayed period. Note that the output of the timer is originally HIGH, thus maintaining the TMR1 contact closed. When the STOP MOTOR1 pushbutton (wired NC) is depressed, while the motor is running, the internal output is energized, which enables the OFF-delay timer. When the timer times out, the contacts open, and the motor is de-energized.

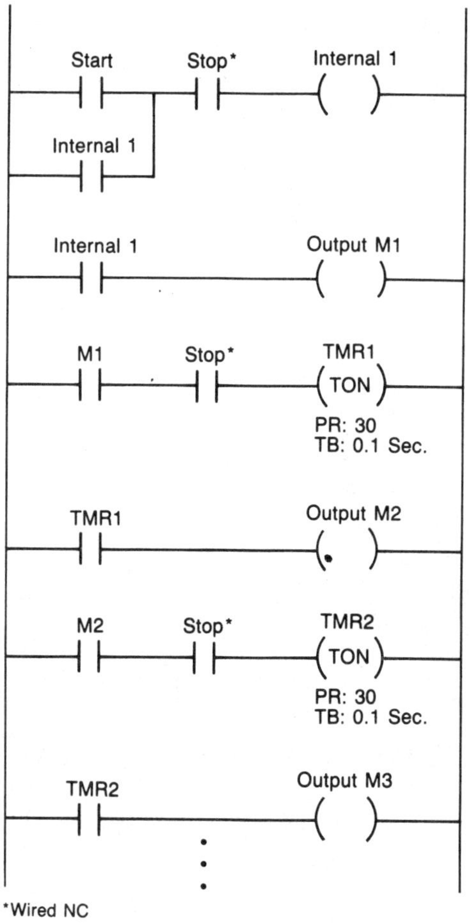

**Figure 8-28.** Sequential motor-start logic.

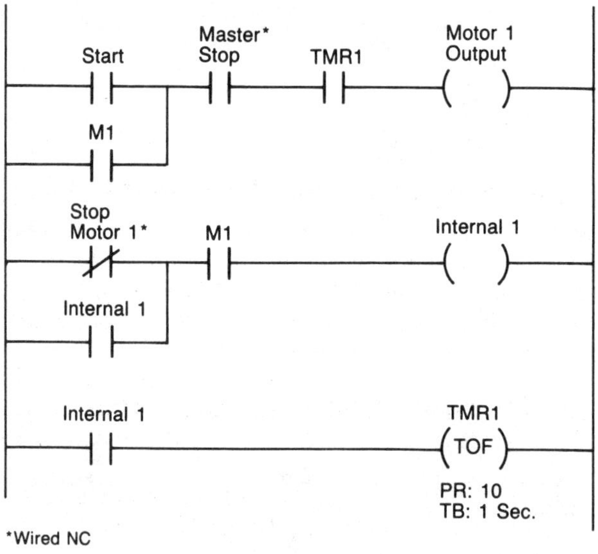

**Figure 8-29.** Delayed de-energize circuit.

## EXAMPLE 12: A 24 Hour Clock

The 24 hour clock has many applications, but is generally used to display the time of day or to determine the time that a report is generated. The logic used to implement the clock is shown in Fig. 8-30. It consists of three counters: one counts 60 seconds, another counts 60 minutes, and the third counts 24 hours. The time can be displayed by outputting the accumulator register values of each counter, to seven-segment BCD displays.

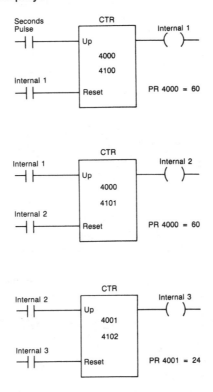

**Figure 8-30.** A 24-hour clock circuit.

## EXAMPLE 13: Counting Beyond The Maximum Count

Depending on the application, it may be necessary to count events that will exceed the maximum allowable number that can be held in a register. The maximum count in most controllers is either 9999 (BCD) or 32767 (binary). To count beyond 9999 simply involves cascading two counters in which the output of the first counter is used as the input of the UP-Count of the second counter. If 32767 is the maximum count, the same approach would result in an erroneous count. The first register would contain a value of 1 (after 32767 is reached), and the second register would contain 00000. This result would indicate a count of 100000 instead of the actual count of 32768.

A solution to this situation is to set the preset value of the first counter to 9999 and use a second counter to register each time 10000 counts occur. The following sequences in Fig. 8-31 illustrate this technique.

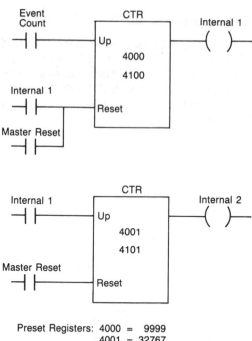

Preset Registers: 4000 = 9999
4001 = 32767

Maximum Count is 327679999
in Accumulated Registers 4100 and 4101

**Figure 8-31.** Counting beyond maximum count.

## EXAMPLE 14: Elimination of Bi-Directional Power Flow

On occasion, when converting relay-logic to program logic, we will find relay circuits that have been designed to allow power flow bi-directionally as shown in Fig. 8-32a. Power can flow down through CR1 or up through CR1 to make a complete path. Bi-directional flow is not allowed in PCs, but the circuit can be restructured to establish a circuit for each direction of power flow. The result, as shown in Fig. 8-32b, is two separate circuits that allow uni-directional, left-to-right power flow.

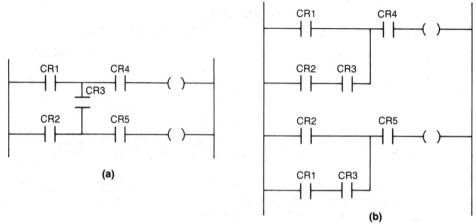

**Figure 8-32.** Circuit with bidirectional power flow **(a)** restructured for PC logic **(b)**.

## EXAMPLE 15: Push-to-Start/Push-to-Stop Circuit

Often, it is desirable to have a single pushbutton perform the start (enable) and stop (disable) functions. In this example (Fig. 8-33) when the pushbutton (PB1) is depressed for the first time, internal output 2 goes HIGH(ON) and remains HIGH. If the pushbutton is depressed again, internal output 2 goes LOW (OFF). The second logic rung detects the first time PB is pressed, while the first rung detects the second time that the button is depressed. A simplified timing diagram shows the operation of this circuit.

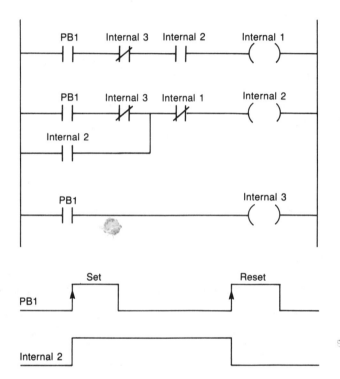

**Figure 8-33.** Push-to-Start/Push-to-Stop circuit.

197

# CHAPTER
# --]9[--

# DOCUMENTING
# THE SYSTEM

*Undocumented changes to a relay control system can be like a bad dream. Undocumented changes to a programmable control system can be a nightmare.*

## 9-1   INTRODUCTION

*Documentation* can be defined as an orderly collection of recorded information, concerning both the operation of a machine or process and the hardware and software components of the control system. These records provide valuable reference for use during system design, installation, start-up, debugging, and maintenance.

To the system designer, documentation should be a working tool that is used throughout the design phase. If the various documentation components are created and kept current during the design, they will not only serve as a reference to the designer, but will also (1) provide an easy means of communicating accurate information to all those involved; (2) put someone else, or the designer, in a position to later answer questions, diagnose possible problems, and modify the program if requirements change; (3) serve as training material for operators who must interface with the system and maintenance personnel who will maintain it; and (4) allow the system to be reproduced easily or altered to serve other purposes.

The achievement of proper documentation is realized through the gathering of hardware, as well as software information. This data is generally supplied from the engineering or electrical group that designs the system and is utilized by the end user. Although documentation is often thought of as something extra, it is a vital system component and should be a general engineering practice. In this chapter, we will cover important aspects needed to provide a "good" PC documentation package that would help in the understanding of the control system.

## 9-2   THE DOCUMENTATION PACKAGE

### System Abstract

As pointed out in the previous chapter, a good design starts with the understanding of the problem and a good description of the process to be controlled. This assessment is followed by a systematic approach that will lead to the implementation of the control system. Once the system is finished, the personnel involved in the design should provide a global description, or abstract, of the scheme and procedure used to control the process.

A system abstract should provide a clear statement of the control problem or task, a description of the design strategy or philosophy used to implement the solution of the problem, and a statement of the objectives that must be achieved. A description of the design strategy should define the function of the major hardware and software components of the system and why they were selected. For instance, a single CPU, located in the warehouse central area, will control two product conveyor lines. A remote subsystem, located in rooms four and five, will control the sorters for those areas. Data will be gathered on total production from both lines and reported in printed form at the end of each shift. Finally, the statement of the objectives will allow the user to measure the success of the control implementation.

The system abstract can always serve to transmit general design information to the end user or anyone who first needs to understand the original control task and, then, reveal the solution to the problem.

### System Configuration

As the name implies, the system configuration is nothing more than a system arrangement diagram. In fact, it is a pictorial drawing of the hardware elements defined in the system abstract. It shows the location, simplified connection, and minimum details regarding major hardware components (i.e. CPU, subsystems, peripherals, etc.). Figure 9-1 illustrates a typical system arrangement diagram.

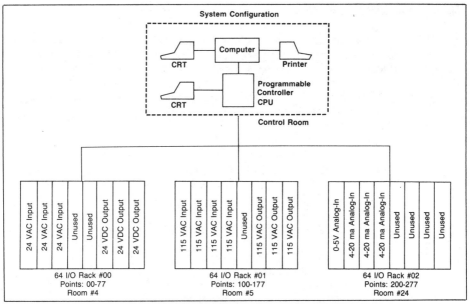

**Figure 9-1.** Typical System Arrangement diagram.

This configuration not only indicates the physical location of subsystems, but also the designation of the I/O rack address assignments. Referencing of the rack address assignments will allow quick location of specific I/O. For example, during start-up the user can easily find out that I/O point 0200 (LS, PB, etc.), located in subsystem 02, is housed in room number four.

## I/O Wiring Connection Diagram

The I/O wiring diagram shows actual connections of the field input and output devices to the PC module. This drawing normally includes power supplies and subsystem connections to the CPU. Figure 9-2 illustrates an example of I/O wiring diagram documentation. The rack, group, and module locations are shown to illustrate explicitly the termination address of each I/O point. If the field devices are not wired directly to the I/O module, then terminal block numbers should be shown. In this example, the terminal blocks are represented by a dark circle with the TB number. Good I/O wiring documentation will be invaluable during installation and for later reference.

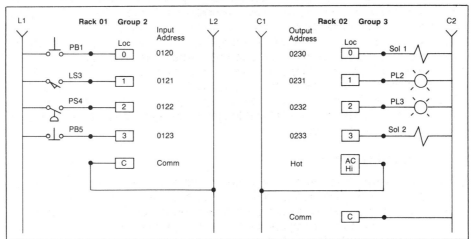

**Figure 9-2.** Typical I/O wiring documents; **(a)** inputs, **(b)** outputs.

## I/O Address Assignments

The I/O Address assignment document identifies each field device by address (based on rack, group, and terminal), the type of input or output module (115 VAC), and the function this device performs in the field. Table 9-1 shows a typical documentation of I/O address assignment.

As can be seen, the I/O address assignment shown in Table 9-1 is similar to the I/O assignment table that was done prior to developing the control program (see Chapter 8).

**Table 9-1.** I/O address assignment document.

| Address | I/O Type | Device | Function |
|---------|----------|--------|----------|
| 0120 | 115 VAC In | PB | Start Push Button PB 1 |
| 0121 | 115 VAC In | LS | Up Limit # 2 |
| 0122 | 115 VAC In | PS | Hydraulic Pressure OK |
| 0123 | 115 VAC In | PB (NC) | Reset PB 2 |
| • | • | • | • |
| • | • | • | • |
| • | • | • | • |
| 0230 | 24 VAC Out | Sol | Retract # 1 |
| 0231 | 24 VAC Out | PL | # 2 in position |
| 0232 | 24 VAC Out | PL | Running |
| 0233 | 24 VAC Out | Sol | Fast Up # 3 |

## Internal I/O Address Assignments

The documentation of internals is an important part of the total documentation and package. Because internals are used for programming timers, counters, and control relay replacement, and are associated with no field devices, there is a tendency to use them freely without much regard for acounting for their usage. However, just as with real I/O, misuse of internals could result in system misoperation.

Good documentation of internals will simplify field modifications during start-up. For example, imagine a start-up situation, involving the modifications of one or more rungs by adding extra interlocking. The user will have to utilize internal coils that are not already being used. If the internal I/O address assignment is current and accurate, showing both used and unused addresses, then a useable address is quickly locatable, time is saved, and confusion is avoided. Table 9-2 illustrates a typical I/O address assignment documentation for internals.

**Table 9-2.** Internal address assignment document.

| Internal | Type or | Description |
|----------|---------|-------------|
| 1000 | Coil | Used to Latch Position |
| 1001 | Coil | Set-up Instantaneous Timer Contact |
| 1002 | Compare | Used for CMP Equal |
| 1003 | Add | Addition Positive |
| • | • | • |
| • | • | • |
| • | • | • |
| T100 | Timer | Time on delay — Motor 1 |
| C400 | Counter | Count pieces on Conv. #1 |
| • | • | • |
| • | • | • |
| • | • | • |

## Register Assignments

Each available system register, whether a user storage register or I/O register, should be properly identified. Most applications use registers to store or hold information for timers, counters, or comparisons. Keeping an accurate record of use and changes to these registers is very critical. Just as with the I/O assignment documents, the register assignment table should show whether an address is being used or not. Table 9-3 shows typical documentation form for register assignments.

**Table 9-3.** Register address assignment document.

| Register | Contents | Description |
|----------|----------|-------------|
| 3036 | Temp In | I/O Register with Analog Module |
| 3040 | Temp In | I/O Register with Analog Module |
| 4000 | 1200 | 20 sec preset of TDR3 |
| 4001 | 2000 | Count preset for CMP = |
| 4002 | 5000 | Count preset for CMP > |
| • | • | |
| • | • | |
| • | • | |
| 4100 | 0 | Not Used |
| to | • | • |
| 4200 | • | • |
| | 0 | • |

## Program Coding Printout

The program .printout is a hard copy of the control logic program stored in the controller's memory. Whether stored in ladder form or some other language, the hardcopy should be an exact replica of what is in memory. Figure 9-3 shows a typical ladder printout.

Generally, a hardcopy printout shows each programmed instruction with the associated address of each input and output. However, information on what each instruction does or what field device is being evaluated or controlled is not readily apparent. For this very reason (lack of complete information), the program coding alone is not adequate for interpretation of the control· system, without the previously mentioned documentation.

The controller will always have the latest software revision of the program stored in memory; therefore, it is wise to have the most recent hardcopy. During start-up, frequent changes are made to the program, which should be immediately documented. It is also a very good practice to obtain the latest hardcopy of the program at the earliest convenience.

## Reproducible Stored Programs

For the most part, PC programming takes place at a different location from where the controller will finally be installed. Saving the control program on a storage medium such as cassette tape, floppy disk, an electronic memory module, or some other means, is always recommended. This practice will allow you to send or carry the stored program to the installation site and reload the controller memory quickly. This approach is usually taken when the system uses a volatile type memory, and even when non-volatile memory is used.

The reproducible stored program, like any other form of documentation, should be kept accurate and current. A good safe practice is to keep two copies always, in case one is damaged or misplaced. Make sure that the stored program coincides with the latest hardcopy of the control logic.

```
RUNG0001
!I0001  I0001  I0001  R0009                                                                MCR
+--] [-+--]/[-+--]/[-+--]/[-+------+------+------+------+------+------+--( )-
!  -01    -02    -03 !  -01
!R0009           '         !
+--]/[-+------+------+     +       +      +       +      +      +      +
!  -02

RUNG0002
 I0001  I0001  R0009  R0010  I0001  R0015  I0001  N0002  N0002  N0004  R0009
+--] [-+--] [-+--]/[-+--]/[-+--]/[-+--] [-+--] [-+--]/[-+--]/[-+--]/[-+--(L)-
!  -02 !  -03    -01 !  -02 !  -05 !  -03 !  -04 !  -01 !  -03 !  -07    -01
!      !I0001            !      !      !I0001 !      !      !      !
+      +--] [-+-SHORT+    +      +--] [-+      +      +      +      +
!         -05            !      !  -06 !
!R0012                   !      !R0013  N0005 !      !      !
+--] [-+------+------+   +--] [-+--] [-+      +      +      +
!  -03                   !         -03    -02 !      !
!I0001  R0009            !                    !      !      !
+--]/[-+--] [-+------+   +      +      +      +      +      +
!  -06    -09            !                    !      !
!      I0001  R0009  I0001                     !                    !
+-OPEN-+--] [-+--] [-+--]/[-+------+------+------+------+      +      +
!         -07 !  -06    -05                    !      !
!            !R0009                            !      !
+      +      +--] [-+------+------+------+------+      +      +
!              -01                             !      !
!R0009                                         !
+--] [-+------+------+------+------+------+------+------+      +
!  -04

RUNG0003
!I0001                                                                       R0009
+--] [-+------+------+------+------+------+------+------+------+------+--(U)-
!  -08                                                                       -01

RUNG0004
 !+------------------+                        +------------------+
+!IF    S0020 =0000 !+------+------+------+------+-TLET   S0020 =0010 !
 !+------------------+                        +------------------+
```

**Figure 9-3.** Typical ladder program printout.

## 9-3  DOCUMENTATION SYSTEMS: AN ALTERNATIVE APPROACH

It is obvious by now that documentation plays a very important role in the design of any programmable controller based system. This documentation may seem, sometimes, to be a tedious and costly task requiring perhaps several knowledgeable people to implement drafting, table preparations, or I/O assignments. As an alternative to this procedure, several manufacturers in the PC support industry have developed sophisticated, yet simple, means of documenting a total programmable controller system.

These systems speed-up the documentation procedure and reduce the manpower needed for the task. They increase the total program development productivity by reducing programming slip-ups and increasing documentation throughput. In addition to the standard types of documentation previously discussed, documentation systems normally provide several other useful documents. A widely used documentation system is the Xycom model 4820 LDT (Ladder Diagram Translator). The following list is a summary of the program documentation features it provides:

- Program titles
- Multiple subtitles
- Date and time the documentation was last produced
- Page numbering
- Extensive commentaries before and after each rung
- Contact or element description
- PC address for each contact
- Pictorial representation of each PC instruction (coils, contacts, etc.)
- Rung numbers
- Rungs where each contact is used
- All preset values of registers used
- Identification of internals and real I/O

Besides the ladder printout, input and output usage (assignment) reports are also provided. These reports show a listing of the controller I/O addresses illustrating how each point is used, whether it is internal or real. Undefined as well as unreferenced symbols are also specified in the report. Complete data table listings and full cross reference reports provide direct information on all register contents and where in the program each element is used. An important advantage of program listings provided by documentation systems is that they show, on a single document, all the information regarding the control program. Figure 9-4 illustrates a typical print out from the Xycom 4820 documentation system.

In general, these systems are capable of uploading, verifying, and storing the PC program directly from the controller or from a cassette, floppy disk, or other storage media.

## 9-4  SUMMARY

Much could still be said for documentation and its relevance to the total system package. The importance of establishing complete, accurate documentation of the control problem and its solution, from the outset, cannot be overemphasized. If the system documentation is created during the system design phases, as it should be, it will not become an unwelcome burden imposed on designers as the project nears completion.

Documentation may seem trivial to some readers or too much work to others. Whether designing their own control system or sub-contracting the design, users should be sure that a good documentation package is delivered with the equipment. A well designed system is one that is not just put to work during start-up, but also one that can be maintained, expanded, modified, and kept running without difficulty. Good documentation will definitely help in this regard for both, the designers and the end user. Remember, regardless of the application, a good design is not good, unless its documentation is also good.

```
11-Feb-83        SAMPLE PRINT FROM THE XYCOM 4820 LDT - RECIPROCATING MOTION MACHINE USING A MODICON 584        Page 1
00:00:47                          THIS PAGE CONTAINS CONVEYOR MOTION CONTROL LOGIC
-----------------------------------------------------------------------------------------------------------------------
    :                                                                                                              :
    :                                                                                                              :
    :                THE RUNG BELOW IS THE START/STOP CIRCUIT FOR THE RECIPROCATING MOTION CONVEYOR               :
    :                                                                                                              :
    :                      +--------------------------------------------------------+                             :
    :                      :  NOTE : STOP PUSHBUTTON IS WIRED NORMALLY CLOSED!!  :                                 :
    :                      +--------------------------------------------------------+                             :
    :                                                                                                              :
    : STOP      START                                                                                             :
    : PUSH      PUSH     WATCH DOG                                                                         : CYCLE
    : BUTTON    BUTTON   SHUT OFF                                                                          : START
    : 10113     10112    00300                                                                  00132      : ] [ 1,2,4
  1 +...] [...+...] [...+---]/[------------------------------------------------------------------( )---+ ]^[ 4
    :           :       :                                                                                :
    :           : CYCLE :                                                                                :
    :           : START :                                                                                :
    :           : 00132 :                                                                                :
    :           +---] [--+                                                                               :
    :                                                                                                    :
    :                                                                                                    :
    :                                                                                                    :
    :                          THE RUNG BELOW LETS THE CONVEYOR STOP ONLY WHEN                           :
    :                          THE PART IS AT THE RIGHT SIDE OF THE MACHINE                              :
    :                                                                                                    :
    :                                                                                                    : CONVEYOR
    : CYCLE                                                                                              : STOP
    : START                                                                                             : CONTROL
    : 00132                                                                                 00133        : ] [ 3
  2 +---] [---+-------------------------------------------------------------------------------( )----+
    :         :
    :CONVEYOR :
    : MOTOR   :
    : FORWARD :
    : 00214   :
    +...] [...+
    :                                                                                                    :
    :LEFT END  CONVEYOR CONVEYOR                                                                         : CONVEYOR
    :OF TRAVEL MOTOR    STOP                                                                             : MOTOR
    :LIMIT SW. REVERSE  CONTROL                                                                          : FORWARD
    : 10114    00215    00133                                                               00214        : ] [ 2
  3 +...] [...-...]/[...---] [------------------------------------------------------...( )...+ ]/[ 3
    :RIGHT END CONVEYOR                                                                                  : CONVEYOR
    :OF TRAVEL MOTOR                                                                                     : MOTOR
    :LIMIT SW. FORWARD                                                                                   : REVERSE
    : 10115    00214                                                                        00215        : ]/[ 3
    +...] [...-...]/[...------------------------------------------------------...( )...+
    :
    :                      THIS TIMING CIRCUITRY KEEPS THE MACHINE CYCLE TIME
    :                      AND AUTOMATICALLY SHUTS OFF THE MACHINE IF LEFT UNATTENDED
    :              CYCLE
    : CYCLE     WATCH DOG                                                                                : WATCH DOG
    : START     TIMER                                                                                    : SHUT OFF
    : 00132     +----+                                                                      00300        : ]/[ 1
  4 +---] [-----:30001:--------------------------------------------------------------------( )---+
    :           :    :
    : STOP      :    :
    : PUSH      :    :
    : BUTTON    :    :
    : 10113     : T1.0:
    +...]/[...--:40001:-
    :           +----+
    :
```

**Figure 9-4(a).** Ladder printout from documentation system (Courtesy of Xycom).

## PROGRAM INFORMATION

```
PC program name ........... : DEMOMOD
Description .............. :
Date/time stored on disk .. : 03/18/82  at 00:00:23
PC type ................... : Modicon 584
Program size .............. : 42
Register memory size ...... : 200
```

## LISTING INFORMATION

```
Time at which listing finished ......... : 00:03:22
Ladder listing uses .................... : network numbers
Cross reference data ................... : shown to right
Closest control for each element ....... : not shown
Maximum number of elements per line .... : 11
Maximum cross references listed per rung : 0
Implied cross references ............... : included
Maximum lines per network .............. : no
Lines per page ......................... : 60
Register listing format ................ : default
```

## REPORT TABLE OF CONTENTS

```
Ladder diagram ......................... : 1
Address usage report ................... : 4
Unreferenced description report ........ : 7
Undefined description report ........... : 8
Full cross reference report ............ : 9
Register memory listing ................ : 10
Traffic cop report ..................... : 11
Configuration report ................... : 12
Message report ......................... : 13
```

## LADDER LISTING SYMBOL KEY

```
     !27 character                        ! output
     !contact                             ! coil
     !description                         ! description
     ! aaaaa      aaaaa            aaaaa  !
bbbb +---]/[---------...] [...-----------------------( )---+ optional
     !ccc                                 ! cross
     !                                    ! reference


     aaaaa =  Contact address
     bbbb  =  Line or rung number
     ccc   =  Closest line or rung number where this
              contact is an output coil
     ...   =  Marks a contact that is a real I/O address
```

## ADDRESS USAGE REPORT SYMBOL KEY

```
     .  =  Address not used in program and not assigned as real I/O
     S  =  Address used in program but not assigned as real I/O
     0  =  Address not used in program but is assigned as real I/O
     *  =  Address used in program and assigned as real I/O
```

**Figure 9-4(b).** 4820 Ladder Translator features (Courtesy of Xycom).

# PART
# --]IV[--

# SPECIAL
# TOPICS

This part consists of subjects related to hardware, software, applications, and PC system analysis. In general, the chapters are independent and can be read in any sequence.

# CHAPTER
# --]10[--

# INSTALLATION, START-UP, AND MAINTENANCE

*A conscientious approach to planning the placement and interconnection of the controller components will simplify installation, start-up, and maintenance and will also insure the best possible machine operation.*

## 10-1   SYSTEM LAYOUT

The design nature of programmable controllers includes a number of rugged design features that allow the PC to be installed in almost any industrial environment. However, a little foresight during the installation will go a long way toward assuring proper system operation.

The system layout is a conscientious approach in placing and interconnecting the components not only to satisfy the application, but also to insure that the controller will operate troublefree in the environment in which it will be placed. With a carefully constructed layout, components are accessible and easily maintained.

In addition to the programmable controller equipment, the system layout also takes into account other components that form part of the total system. These other types of equipment include isolation transformers, auxiliary power supplies, safety control relays, and incoming line noise suppressors.

Although it is not necessary to install the controller in a controlled environment because it is designed to work on the factory floor, certain considerations should be followed for the proper installation of the system components. The best place for the PC to be located is near the machine or process that it will control, if temperature, humidity, and electrical noise are not problems. Placing the controller near the equipment and using remote I/O where possible will minimize wire runs and simplify start-up and maintenance. Figure 10-1 shows a PC installation.

### Enclosures

PCs are generally placed in a NEMA-12 or JIC enclosure. The enclosure size depends on the total space required. Mounting the controller components in an enclosure is not always required, but is recommended for most applications to protect against atmospheric contaminants, such as conductive dust, moisture, and any other corrosive or harmful airborne substances. Metal enclosures also help to minimize the effects of electromagnetic radiation that may be generated by surrounding equipment.

### Enclosure Layout

The enclosure layout should conform to NEMA standards, and placement and wiring of the components should take into consideration the effects of heat, electrical noise, vibration, maintenance, and safety. A typical enclosure layout is illustrated in Fig. 10-2 and can be used for reference with the following layout discussions.

**General.** The following recommendations are concerned with preliminary considerations for the location and physical aspects of the enclosure.

- The enclosure should be placed in a position that allows the doors to be opened fully for easy access to wiring and components for testing or trouble-shooting.

- The enclosure depth should be enough to allow clearance between a closed enclosure door and any print pocket mounted on the door or the enclosed components and related cables.

- The enclosure's back panel should be removable to facilitate mounting of the components and other assemblies.

- An emergency disconnect device should be mounted in the cabinet in an easily accessible location.

- Accessories, such as AC power outlets, interior lighting, or a gasketed plexiglass window to allow viewing of the processor or I/O indicators should be considered for installation and maintenance convenience.

**Figure 10-1.** Installation of a PC at a central location **(a),** and installation of a PC close to the controlled machine **(b)** (Bivans' vertical cartoner machine). Courtesy Omron Electronics and Bivans.

**Environmental.** The effects of temperature, humidity, electrical noise, and vibration are important factors that will influence the actual placement of the controller, the inside layout of the enclosure, and the necessity for other special equipment. The following points are concerned with insuring favorable environmental conditions for the controller.

- The temperature inside the enclosure must not exceed the maximum operating temperature of the controller (typically 60 C max.).

- If "hot spots" such as those generated by power supplies or other electrical equipment are present, a fan or blower should be installed to help dissipate the heat.

- If condensation is anticipated, a thermostatically controlled heater may be installed inside the enclosure.

213

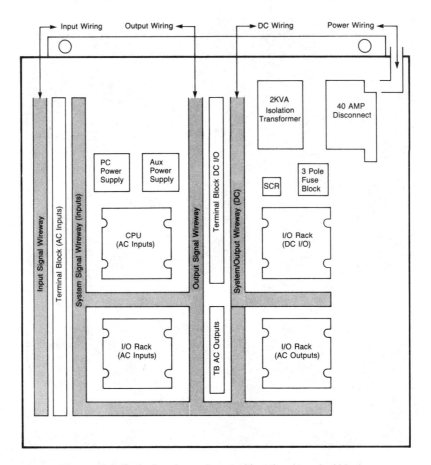

**Figure 10-2.** Typical enclosure layout with optional terminal blocks.

- If the area in which the system will be located contains equipment that generates excessive electromagnetic interference (EMI) or radio frequency interference (RFI), the enclosure should be placed well away from these sources. Examples of such equipment are welding machines, induction heating equipment, and large motor starters.
- In cases in which the PC enclosure must be mounted on the controlled equipment, careful attention should be given to insure that vibrations caused by that equipment do not exceed the PC's vibration specifications.

**Placement of PC Components .** The placement of the major components of a specific controller is dependent on the number of system components and the physical design or modularity of each component (see Fig. 10-3). Although various controllers may have different mounting and spacing requirements, the following considerations and precautions are applicable when placing any PC inside an enclosure.

- To allow maximum convection cooling, all controller components should be mounted in a vertical (upright) position. Some manufacturers may specify that the controller components can be mounted horizontally. However, in most cases, components mounted horizontally will obstruct air flow.

**Figure 10-3.** Installation of a TI model 530 PC with over 600 I/O (Courtesy of Texas Instruments Corp.).

- The power supply (main or auxiliary) has a higher heat dissipation than any other system component and, therefore, should not be mounted directly underneath any other equipment. Typically, the power supply is installed above all other equipment and at the top of the enclosure with adequate spacing (at least ten inches) between the power supply and the top of the enclosure. The power supply may also be placed adjacent to other components, but with sufficient spacing.

- The CPU should be placed at a comfortable working level (e.g. at sitting or standing eye level) that is either adjacent to or below the power supply. If the CPU and power supply are contained in a single unit, then it should be placed toward the top of the enclosure with no other components directly above it, unless there is sufficient spacing.

- Local I/O racks (same panel enclosure as CPU) are placed in any desired arrangement within the distance allowed by the I/O rack interconnection cable. Typically the racks are placed below or adjacent to the CPU, but not directly above the CPU or power supply.

- Remote I/O racks (away from the CPU) and their auxiliary power supply are generally placed inside an enclosure at the remote location. The same placement practices described for local racks can be followed.

- Spacing of the controller components to allow proper heat dissipation should adhere to the manufacturer's specifications for vertical and horizontal spacing between major components.

**Placement of Other Components.** In general, the placement of other components inside the enclosure should be away from the controller components, so as to minimize the effects of noise or heat generated by these devices. The following list outlines some common practices for locating other equipment inside the enclosure.

215

- Incoming line devices, such as isolation or constant voltage transformers, local power disconnects, and surge suppressors, are normally located near the top of the enclosure and alongside the power supply. This placement assumes that the incoming power enters at the top of the panel. The proper location of incoming line devices keeps power wire runs as short as possible, thus minimizing the chances of transmitting electrical noise to the controller components.

- Magnetic starters, contactors, relays, and other electro-mechanical components should be mounted near the top of the enclosure in an area segregated from the controller components. A good practice is to place a barrier with at least six inches separation between the magnetic area and the controller area. Typically, magnetic components are placed adjacent and opposite to the power supply or incoming line devices.

- If fans or blowers are used for cooling the components inside the enclosure, they should be placed in a location close to the heat generating devices (generally power supply heat sinks). When using fans, be sure that outside air is not brought inside the enclosure unless a fabric or other reliable filter is also used. This filtration will prevent conductive particles or other harmful contaminants from entering the enclosure.

**Grouping Common I/O Modules.** The grouping of the I/O will allow signal and power lines to be routed properly through the ducts, such that crosstalk interference will be minimized. The following are recommendations for grouping I/O modules.

- I/O modules should be segregated into groups such as AC input modules, AC output modules, DC input modules, DC output modules, analog input modules, and analog output modules whenever possible.

- If possible, a separate I/O rack should be reserved for common input or output modules. If it is not possible to reserve a complete rack for common modules, then the modules should be separated as much as possible within a rack. As an example, a suitable partitioning would involve placing all AC together or all DC together, and if space permits, allowing an unused slot between the different groups.

**Duct and Wiring Layout.** The duct and wiring layout defines the physical location of wireways and the routing of field I/O signals, power, and controller interconnections within the enclosure. The enclosure's duct and wiring layout is dependent on the placement of I/O modules within each I/O rack (see Grouping Common I/O). The placement of these modules is determined during the design stages, when the I/O assignment takes place (see Chapter 8). Prior to defining the duct and wiring layout and assigning the I/O, the following practices should be considered to minimize electrical noise caused by crosstalk between I/O lines.

- All incoming AC power lines should be kept separate from low-level DC lines, I/O power supply cables, and all I/O rack interconnect cables.

- Low-level DC I/O lines, such as TTL and analog, should not be routed in the same duct in parallel with AC I/O lines. Whenever possible, always keep AC signals separate from DC signals.

- I/O rack interconnect cables and I/O power cables can be routed together in a common duct not shared by other wiring. When this arrangement is impractical or it is impossible to separate these cables from all other wiring, then they can be routed with low-level DC lines or routed externally to all ducts and held in place using tie wraps or some other fastening method.

216

- If I/O wiring must cross the AC power lines, it should do so only at right angles. This routing practice will minimize the possibility of electrical noise pick-up.
- When designing the duct layout, at least two inches should be allowed between the I/O modules and any wire duct. If terminal strips are used, then two inches should be allowed between the terminal strip and wire duct as well as between the terminal strip and I/O modules.

**Grounding.** Proper grounding is an important safety measure in all electrical installations. When installing electrical equipment, users should refer to the National Electric Code (NEC), article 250, which provides data such as the size and types of conductors, color codes, and connections necessary for safe grounding of electrical components. The code specifies that a grounding path must be permanent (no solder), continuous, and able to conduct safely the ground-fault current in the system with minimal impedance. The following grounding practices will have significant impact on the reduction of noise caused by electromagnetic induction.

- Ground wires should be separated from the power wiring at the point of entry to the enclosure. To minimize the ground wire length within the enclosure, the ground reference point should be located as close as possible to the point of entry of the plant power supply.
- All electrical racks or chassis and machine elements should be grounded to a central ground bus, normally located in the magnetic area of the enclosure. Paint or other non-conductive materials should be scraped away from the area where a chassis makes contact with the enclosure. In addition to the ground connection made through the mounting bolt or stud, a one inch metal braid or size #8 AWG wire (or manufacturer's recommendation) can be used to connect between each chassis and the enclosure at the mounting bolt or stud.
- The enclosure should be properly grounded to the ground bus; insure that good electrical connection is made at the point of contact with the enclosure.
- The machine ground should be connected to the enclosure and to earth ground.

## System Power

PC power supplies operate, in general, with sources of 120 or 240 VAC single phase. If the controller is installed in an enclosure, the two power leads (L1 hot, L2 common) normally enter the enclosure through the top part of the cabinet to minimize interference with other control lines. It is important to supply the cleanest possible power line so that problems related with line interference to the controller and I/O system are avoided.

**Common AC Source.** It is always good practice to use a common AC source to the system power supply and the I/O devices. This practice will minimize line interference and prevent the possibility of reading faulty input signals if the AC source to the power supply and CPU is stable, but the AC source to the I/O devices is unstable. By keeping both the power supply and I/O devices on the same power source, the user can take full advantage of the power supply's own line monitoring feature. If line conditions fall below the minimum operating level, the power supply will detect the abnormal condition and signal the processor, which will stop reading input data and turn off all outputs.

**Isolation Transformers.** Using isolation transformers on the incoming AC power line to the controller is another good practice. An isolation transformer is especially desirable in cases in which heavy equipment will be likely to introduce noise onto the AC line. The isolation transformer can also serve as a step-down transformer to reduce the incoming line voltage to a desired level. The transformer should have a sufficient power rating (units of volt-amperes) to supply the load adequately. Users should consult the manufacturer for the recommended transformer rating for a particular application.

## Safety Circuitry

Sufficient emergency circuits should be provided to stop either partially or totally the operation of the controller or the controlled machine or process. These circuits should be routed outside the controller, so in the event of total controller failure, independent and rapid shutdown means are available. Devices, such as emergency pull rope switches or end of travel limit switches, should operate motor starters, solenoids, or other devices without being processed by the controller. These emergency circuits should be implemented using simple logic with a minimum number of highly reliable, preferably electromechanical, components.

**Emergency Stops.** It is recommended that emergency stop circuits be incorporated into the system for every machine directly controlled by the PC. To provide maximum safety in programmable controller systems, these circuits must not be wired into the controller, but should be left hardwired. Emergency stop switches should be placed in locations that are easily accessible to the operator and there should be as many as are required. These emergency stop switches are generally wired into master control relay (MCR) or safety control relay (SCR) circuits that will remove power from the I/O system in an emergency.

**Master or Safety Control Relays.** MCRs or SCRs provide a convenient means for removing power from the I/O system during an emergency situation. By de-energizing the MCR (or SCR) coil, power to the input and output devices is removed. This event occurs when any emergency stop switch opens. However, the CPU continues to receive power and to operate even though all its inputs and outputs are disabled.

The MCR circuit may be extended by placing the PC fault relay (closed during normal PC operation) in series with any other emergency stop conditions. This enhancement will cause the MCR circuit to drop the I/O power in case of a PC failure (memory error, I/O communications error, etc.). Figure 10-4 illustrates typical wiring of safety circuits.

**Emergency Power Disconnect.** A properly rated emergency power disconnect should be used in the power circuit feeding the power supply as a means of removing power from the entire programmable controller system (see Fig. 10-4). A capacitor (0.47 uF for 120 VAC, 0.22 uF for 220 VAC) is sometimes placed across the disconnect to protect against a condition known as *"outrush."* Outrush occurs when the output Triacs are turned off by throwing the power disconnect, thus causing the energy stored in the inductive loads to seek the nearest path to ground, which is often through the Triacs.

## Special Considerations

If the recommendations that have been previously outlined are closely followed, they should provide favorable operating conditions for most programmable controller applications. However, in certain applications the operating environment may not be considered normal and may have extreme conditions that require special attention. These adverse conditions include excessive noise and heat and nuisance line fluctuations. The following discussions describe these conditions and possible measures that will minimize their effects.

**Excessive Noise.** Electrical noise is seldom responsible for the damaging of components, unless extremely high energy or high voltage levels are present. However, the malfunctions due to noise are temporary occurrences of operating errors that can result in hazardous machine operation in certain applications.

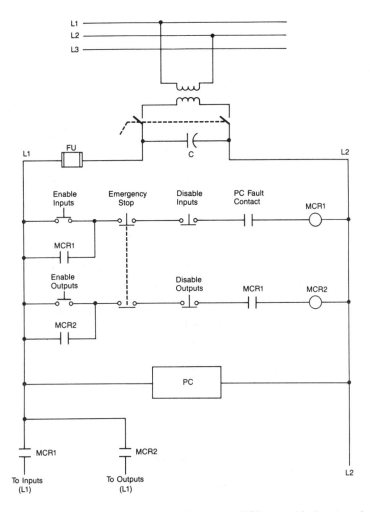

**Figure 10-4.** Typical safety wiring diagram with separate MCR control for inputs and outputs.

Noise may be present only at certain times, may appear at widely-spread intervals, or in some cases may exists continuously. The second case is the most difficult to isolate and to correct.

Noise usually enters through input, output, and power supply lines and may be coupled into the lines electrostatically through the capacitance between these lines and noise signal carrier lines. This effect generally results from the presence of high voltage or long, closely-spaced conductors. Coupling of magnetic fields can also occur when control lines are closely spaced with lines carrying large currents. Devices that are potential noise generators include relays, solenoids, motors, and motor starters, especially when operated by hard contacts such as pushbuttons or selector switches.

Although a great deal of effort has been placed in the design of solid-state controls to achieve reasonable noise immunity, special considerations must be given to minimize noise, especially when the anticipated noise signal has characteristics similar to the desired control input signals. To increase the operating noise margin, the controller must be located away from noise generating devices such as large AC motors and high-frequency welding machines. All inductive loads

219

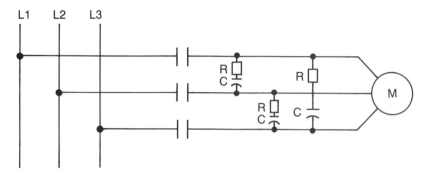

**Figure 10-5.** Example of three-phase motor leads suppression.

must be suppressed. Three-phase motor leads should be grouped together and routed separately from low-level signal leads. If the situation is critical, it may be necessary to suppress all three-phase motor leads, as shown in Fig. 10-5. Suppression techniques for small and large inductive devices are described in the next section.

**Excessive Heat.** Programmable controllers are designed to withstand temperatures in the range of 0 to 60 °C and are normally cooled by convection, in which a vertical column of air is drawn in an upward direction over the surface of the components. To keep the temperature within limits, the cooling air at the base of the system must not exceed 60 °C. Based on these specifications, proper spacing must be allocated between components when they are installed. The spacing recommendations from manufacturers are, in general, determined for typical conditions which exist for most applications. Conditions are considered typical if, usually, 60% of the inputs are ON at any one time, 30% of the outputs are ON at any one time, current supplied by all the modules combined average a certain value, and the air temperature is around 40 °C.

Situations in which most of the I/O are ON at the same time and the air temperature is higher than 40 °C are not typical. Spacing between components must be larger to provide better convection cooling. If equipment inside or outside of the enclosure generates substantial amounts of heat and the I/O system is switched ON continuously, a fan should be utilized inside the enclosure to provide good air circulation to reduce hot spots created near the PC system. When using a fan, the air being brought in should first pass through a filter to prevent dirt or other contaminants from entering. Dust will obstruct the heat dissipation from components and can be especially harmful on heat sinks when thermal conductivity to the surrounding air is lowered. In cases of extreme heat, an air conditioning unit should be used to prevent heat build-up inside the enclosure.

**Excessive Line Voltage Variation.** The power supply section of the PC system is built to sustain line fluctuations and still allow the system to function within its operating margin. As long as the incoming voltage is adequate, the power supply will continue to provide all the logic voltages necessary to support the processor, memory, and I/O. If the voltage drops below the minimum acceptable level, the power supply will signal the processor and a system shut-down will be executed.

In cases in which the installation is subject to *"soft"* AC lines and unusual line variations, a constant voltage transformer can be used to prevent the system from shutting-down too often. However, a first step toward the solution of the line variations is to correct any possible feeder problem in the distribution system. If this correction does not solve the problem, a constant voltage transformer must be used.

The constant voltage transformer stabilizes the input voltage to the power supply and input field devices by compensating for voltage changes at the primary in order to maintain a steady voltage at the secondary. When using a constant voltage transformer, the user should be sure to check that its power rating is sufficient to supply the input devices and the power supply. The output devices are generally connected to the line in front of the constant voltage transformer, instead of providing power to the outputs from the transformer. This arrangement will lessen the load supported by the transformer and allow a smaller rating. The information regarding power rating requirements can be obtained from the manufacturer.

## 10-2    INPUT/OUTPUT INSTALLATION

The input/output installation is perhaps the biggest and most critical job when it comes to installing the programmable controller system. To minimize errors and simplify installation, predefined guidelines must be followed. The guidelines for installing the I/O system should have been prepared during the design phase and should be provided to the persons involved in the controller installation. A complete set of documents with precise information regarding I/O placement and connections will assure that the intended total system organization is achieved. Furthermore, these documents should be constantly updated during every stage of the installation. The following considerations will facilitate an orderly installation.

### Preliminary Wiring Considerations

**Wire Size.** Each I/O terminal has been designed to accept one or more conductors of a particular wire gauge. This wire size should be checked to insure that the correct gauge is being used and that it is properly sized to handle the maximum possible current.

**Wire and Terminal Labeling.** A label should be used to identify each field wire and its termination point. Reliable labeling methods such as shrink-tubing or tape should be used on each wire, and tape or stick-on labels should be used on each terminal block. Color coding of similar signal characteristics (e.g. AC-red, DC-blue, common-white, etc.) can be used in addition to wire labeling. Typical labeling nomenclature includes wire numbers, device names or numbers, or the input or output address assignment. Good wire and terminal identification will simplify maintenance and trouble-shooting.

**Wire Bundling.** Wire bundling is a technique commonly used to simplify the connections to each I/O module. When using this method, the wires that will be connected to a single module are bundled, generally using a tie-wrap, and then routed through the duct with other bundles of the same signal characteristics. The input, power, and output bundles carrying the same type of signals should be kept in separate ducts when possible to avoid interference.

### I/O Module Placement

Placement and installation of the I/O modules is simply a matter of inserting the correct modules in their proper locations. This procedure involves verifying the type of module (115 VAC input, 24 VDC output, etc.) and the slot address as defined by the I/O address assignment document. Each terminal on the module will be wired to field devices that have been assigned to that termination address. The user should be sure that power to the modules (or rack) is removed before installing and wiring any module.

## Wiring Procedures

Once I/O modules have been placed in the correct slots and wire bundles have been made for each module, the wiring to the modules can take place. The following procedures are recommended for I/O wiring:

- Remove and lock-out input power from the controller and I/O before any installation and wiring begins.
- Verify that all modules are in the correct slots. Check module type and model number by inspection, and I/O wiring diagram. Check the slot location according to the I/O address assignment document.
- Loosen all terminal screws on each I/O module.
- Locate the wire bundle corresponding to each module and route it through the duct to the module location. Identify each of the wires in the bundle and be sure that they correspond to that particular module.
- Starting with the first module, locate from the bundle the wire that connects to the lowest terminal. At the point where the wire is at a vertical height equal to the termination point, bend the wire at a right angle towards the terminal.
- Cut the wire at a length that extends 1/4 inch past the edge of the terminal screw. Strip the insulation from the wire approximately 3/8 of an inch. Insert the uninsulated end of the wire under the pressure plate of the terminal and tighten the screw.
- If two or more modules share the same power source, the power wiring can be jumpered from one module to the next.
- If shielded cable is being used, connect only one end to ground, preferably at the rack chassis. This connection will avoid possible ground loops. The other end should be left cut-back and unconnected, unless specified otherwise.
- Repeat the wiring procedure for each wire in the bundle until the module wiring is complete.
- After all wires are terminated, check for good terminations by gently pulling on each wire.
- Repeat the wiring procedure until all modules are completed.

## Special Connection Considerations

Typical connection diagrams for the various types of I/O modules have been illustrated in Chapter 5. However, certain field device wiring connections may need special attention. These connections include leaky inputs, inductive loads, output fusing, and use of shielded cable.

**Connecting Leaky Inputs.** Some field devices will have a small leakage current even when they are in the OFF state. Both triac and transistor outputs will exhibit this leakage characteristic, although the transistor leakage current is much lower. Often, the leaky input will only cause the module's input indicator to flicker. The leakage could, however, result in misoperation by falsely triggering an input circuit. This situation can be corrected by placing a "bleeding" resistor across the input. A typical device that exhibits this situation is a proximity switch. This leakage may also be observed when driving an input module with an output module when there is no other load. Figure 10-6 illustrates both of these cases along with the corrective action taken.

**Suppression of Inductive Loads.** When the current in an inductive load is interrupted (by turning the output OFF), a very high voltage spike is generated. These spikes, if not suppressed, can reach several thousand volts across the

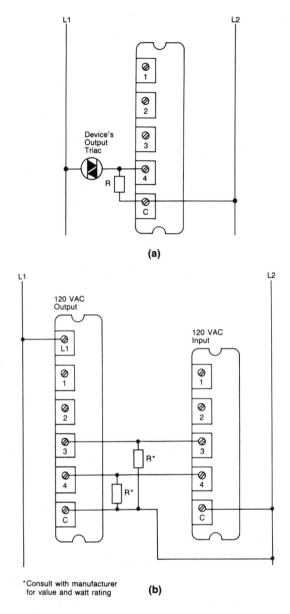

**Figure 10-6. (a)** Typical connection for leaky input devices.
**(b)** Connection of output module to input module.

leads which feed power to the device or between both power leads and chassis ground, depending on the physical construction of the device. This high voltage causes erratic operation and, in some cases, may damage the output module. To avoid this situation, a snubber circuit, typically an RC or MOV, should be installed to limit the voltage spike as well as the rate of change of current through the inductor.

Generally, output modules are designed to drive inductive loads and often include suppression networks; however, under certain loading conditions, it may be found that the triac is unable to turn OFF as current passes through zero (commutation), thus requiring external suppression. If this is the case, additional suppression must be installed.

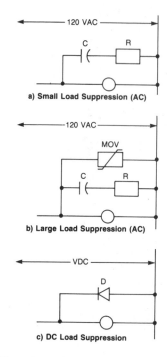

**Figure 10-7.** Example suppression networks for AC and DC loads.

Suppression for small AC devices, such as solenoids, relays, and motor starters up to Size 1, can be accomplished by placing an RC snubber circuit across the device; larger contactors of Size 2 and above will require an MOV in addition to the RC network (see Fig. 10-7). DC suppression is accomplished by using a free-wheeling diode across the load. Figure 10-8 illustrates several examples of suppressing inductive loads.

**Fusing Outputs.** Solid state outputs are normally provided with fusing on the module, to protect the triac or transistor from moderate overloads. If fuses are not provided internally, they should be installed (normally at the terminal block) externally, during the initial installation. When adding fuses to the output circuit, be sure to adhere to the manufacturer's specification for a particular module. Only a properly rated fuse will insure that, in an overload condition, the fuse will open quickly to avoid overheating of the output switching device.

**Shielding.** Control lines such as TTL, analog, thermocouple, and other low-level signals are normally routed in a separate wireway, to reduce the effects of signal coupling. For further protection, shielded cable is used to protect the low-level signals from electrostatic and magnetic coupling with lines carrying 60 Hz power and other lines that carry rapidly changing currents. The twisted, shielded cable should have at least a one inch lay, or approximately twelve twists per foot, and be protected on both ends by means such as shrink tubing. The shield should be connected to control ground at only one point, as shown in Fig. 10-9, and shield continuity must be maintained for the entire length of the cable. Care should be taken to insure that the shielded cable is routed away from high noise areas, is insulated over its entire length, and makes a connection at ONLY one point.

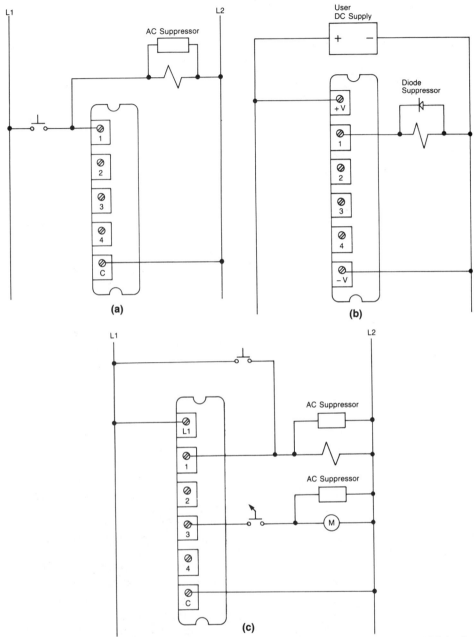

**Figure 10-8.** Suppression of inductive loads in parallel with PC input **(a)**; suppression of DC load **(b)**; and suppression of loads with switch in parallel and series with PC output **(c)**.

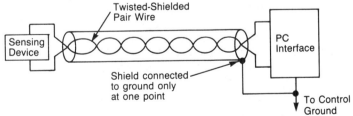

**Figure 10-9.** Example of shielded cable ground connection.

## 10-3   THE SYSTEM START-UP

### Pre-Start-Up Procedures

Prior to applying power to the system, several final inspections of the hardware components and interconnections are recommended. These recommended procedures will undoubtedly require time up front; however, this invested time will almost always insure a reduced total start-up time, especially for large systems with many input/output devices. The following checklist can be used for pre-start-up procedures:

• Visually inspect to insure that all PC hardware components are present. Verify correct model numbers for each component.

• Inspect all CPU and I/O modules to insure that they are installed in the correct slot locations and are placed securely in position.

• Check that the incoming power is wired correctly to the power supply (and transformer) and that the system power is properly routed and connected to each I/O rack.

• Verify that each I/O communication cable that links the processor and each I/O rack is correct, according to the I/O rack address assignment.

• Verify that all I/O wiring connections at the controller end are in place and are securely terminated. This check involves using the I/O address assignment document to verify that a wire is terminated at each point as specified by the assignment document.

• Verify that output wiring connections are in place and properly terminated at the field device end.

• For maximum safety, the system memory should be cleared of any control program that has been previously stored. If the control program is stored in EPROM, then the chips should be removed temporarily.

### Static Input Wiring Check

The static input wiring check is performed with power applied to the controller and input devices. When performed, it will verify that each input device is connected to the correct input terminal and that the input module or point is functioning properly. Since these tests are performed first, they will also verify that the processor and the programming device are in good working condition. Proper input wiring can be verified using the following procedures:

• Place the controller in a mode that will inhibit the PC from any automatic operation. This mode will vary depending on the PC model, but is typically: STOP, DISABLE, PROGRAM, etc.

• Apply power to the system power supply and input devices. Verify that all system diagnostic indicators are indicating proper operation. Typical indicators are: AC OK, DC OK, Processor OK, Memory OK, and I/O Communication OK.

• Verify that the emergency stop circuit will de-energize power to the I/O devices.

• Manually activate each input device and observe the corresponding LED status indicator on the input module and/or monitor the same address on the programming device, if used. If properly wired, the indicator will turn ON. If an indicator other than the expected one turns ON when the input device is activated, then it is possible that the wiring is to the wrong input terminal. If no indicator turns ON, then a fault possibly exists in either the input device, field wiring, or the input module (see Section 10-4).

- Precautions should be taken when activating input devices that are connected in series with loads that are external to the PC and could cause injury or damage.

## Static Output Wiring Check

The static output wiring check is performed with power applied to the controller and output devices. A good safe practice, however, is first to disconnect locally all output devices that will involve mechanical motion (e.g. motors, solenoids, etc.). When performed, the static output wiring check will verify that each output device is connected to the correct terminal address and that the device and output module are functioning properly. The output wiring can be verified using the following procedures:

- Locally disconnect all output devices that will cause mechanical motion.

- Apply power to the controller and to the input/output devices. If power to the outputs can be removed by an emergency stop, verify that the circuit does remove power when activated.

- Static checkout of the outputs should be performed one at a time. If the device is a motor or other output that has been locally disconnected, reapply power to that device only, prior to checking. The output operation inspection can be performed using one of the following methods:

    1) Assuming the controller has a forcing function, each output can be tested with the use of the programming device by forcing the output ON and by setting the corresponding terminal address (point) to 1. If properly wired, the corresponding LED indicator will turn ON and the device will energize. If an indicator other than the expected one turns ON when the terminal address is forced, then it is possible that the wiring is to the wrong output terminal. Inadvertent machine operation is avoided since rotating and other motion producing outputs are disconnected. If no indicator turns ON, then a fault possibly exists in either the output device, field wiring, or the output module (see Section 10-4).

    2) An alternative to forcing involves programming a "dummy" program rung that can be used repeatedly for testing each output. Program a single rung with a single normally-open contact (a conveniently located pushbutton) controlling the output. To test the output, the CPU must be placed in either the RUN or *Single-Scan* mode, or a similar mode, depending on the controller. With the controller in the RUN mode, the test is performed by depressing the pushbutton. If the Single-Scan is used, it is performed by depressing (and maintaining) the pushbutton while the Single-Scan is executed. Observe the output device and LED indicator, as described in the first procedure.

## Pre-Start-Up Program Check

The pre-start-up program checkout is simply a final review of the control program. This check can be performed at any time, but should be done prior to loading the program into memory for the dynamic system checkout.

To perform the check will require a complete documentation package that relates the control program to the actual field devices. Documents such as address assignments and wiring diagrams should reflect any modifications that may have occurred during the static wiring checks. When performed, this final program review will verify that the final hardcopy, which will be loaded into memory, is error-free or at least agrees with the original design documents. The following is a checklist for final program checkout:

- Using the I/O wiring document verify against the hardcopy program printout that every controlled output device has a programmed output rung of the same address.
- Inspect the hardcopy printout for any entry errors that may have occurred while entering the program; verify that all program contacts and internal outputs have valid address assignments.
- Verify that all timer, counter, and other preset values are as were intended.

## Dynamic Checkout

The dynamic checkout is a procedure by which the logic of the control program is verified for correct operation of the outputs. This checkout assumes that all static checks have been performed, the wiring is correct, the hardware components are operational and functioning correctly, and the software has been thoroughly reviewed. At this point, it can be assumed that it is safe to gradually bring the system under full automatic control.

Although it may not be necessary to start up a small system partially it is always a good practice to start large systems in sections. Large systems generally use remote subsystems that control different sections of the machine or process. Bringing one subsystem on-line at a time will allow the total system start-up to be performed with maximum safety and efficiency. Remote subsystems can be temporarily disabled either by locally removing power or by disconnecting the communications link with the CPU. The following practices outline possible procedures for the dynamic system checkout:

- Load the control program into the PC memory.
- The control logic can be tested, in most cases, using one of the following methods:
    1) A mode such as TEST, if available, will allow the control program to be executed and debugged while the outputs are disabled. A check of each rung can be done by observing the status of the output LED indicators or by monitoring the corresponding output rung on the programming device.
    2) If the controller must be in the RUN mode to update outputs during the tests, outputs that are not being tested and could cause damage or harm should be locally disconnected until they are tested. If an MCR or similar instruction is available, it can be used to bypass execution of the outputs that are not being tested, so that disconnection of the output devices is not necessary.
- Check each rung for correct logic operation and modify the logic if necessary. A useful tool for debugging the control logic is the single-scan. This procedure will allow users to observe each rung as every scan is executed under their command.
- When all the logic has proven to control the outputs satisfactorily remove all temporary rungs that may have been used (MCRs, etc.). Place the controller in the RUN mode and test the total system operation. If all procedures have checked correctly, the full automatic control should operate smoothly.
- All modifications to the control logic should be documented immediately and revised on the original documentation. A reproducible copy (e.g. cassette

recording, etc.) of the program should be obtained at the earliest convenience.

The start-up recommendations and practices that have been presented in this section are considered good procedures that will aid in the safe and orderly start-up of any programmable control system. However, depending on the controller that is being installed, there may be specific start-up requirements that are outlined in the manufacturer's product manual. You should be aware of these specific start-up procedures, prior to attempting to start-up the controller.

## 10-4  MAINTENANCE AND TROUBLE-SHOOTING

Although programmable controllers have been designed in such a way to minimize maintenance for troublefree operation, there are several maintenance aspects that should be taken into consideration once the system has been installed and is operational. Certain maintenance measures, if performed periodically, will minimize the chances of system malfunction. This section outlines some of the practices that should be observed to keep the system in good operating condition.

### Preventive Maintenance

Preventive maintenance of programmable controller systems includes only a few basic procedures or checks that will greatly reduce the failure rate of system components. Preventive maintenance for the PC system could be scheduled with the regular machine or equipment maintenance so that the equipment and controller are down for a minimum amount of time. However, depending upon the environment in which the PC is located, the required preventive maintenance may be more frequent than in other surroundings. The following preventive measures should be taken:

- Any filters that have been installed in enclosures should be cleaned or replaced periodically. This practice will insure that clean air circulation is present inside the enclosure. Filter maintenance should not be put-off until the scheduled machine maintenance, but should be checked periodically at a frequency dependent on the amount of dust in the area.

- Dirt and dust should not be allowed to accumulate on the PC components. To allow heat dissipation, the CPU and I/O system are generally not designed to be dust-proof. If dust is allowed to build-up on heat sinks and electronic circuitry, an obstruction of heat dissipation could occur and cause circuit malfunction. Furthermore, if conductive dust reaches the electronic boards, a short circuit could result and cause permanent damage to the circuit board.

- The connections to the I/O modules should be periodically checked to insure that all plugs, sockets, terminal strips, and module connections are making good connections, and that the module is securely installed. This type of check should be done more often in situations in which the PC system is located in areas that experience constant vibration that could loosen terminal connections.

- Care should be taken to insure that heavy noise generating equipment is not moved too closely to the PC.

- The personnel performing the maintenance should make sure that unnecessary articles are kept away from the equipment inside the enclosure. Leaving articles such as drawings, installation manuals, or other booklets on top of the CPU rack or other rack enclosures could obstruct the air flow and create hot spots, which can result in system malfunction.

## Spare Parts

It is a good practice to keep on hand a stock of replacement parts. This practice will minimize any down time resulting from component failure. In a failure situation, having the right spare in stock could mean a shutdown of only minutes, instead of hours or days. As a rule of thumb, the spares stocked should be 10% of the number of each module used. If a module is used infrequently, then less than 10% of that particular module can be stocked.

Main CPU board components should have one spare each, regardless of how many CPUs are being used. Each power supply, whether main or auxiliary, should have a backup. There are certain applications that may require a complete CPU rack as a stand-by spare. This extreme case exists when there is no time during a failure for determining which CPU board has failed, and the system must be brought into operation immediately.

## Module Replacement

If a module has to be replaced, the user should make sure that the module being installed is of the correct type. Some I/O systems allow modules to be replaced while power is still applied, but others may require that power be removed. If replacing a module solves the problem, but the failure reoccurs in a relatively short period, the user should check inductive loads that may be generating voltage and current spikes and may require external suppression. If the module fuse blows again after it is replaced, it may be that the module's output current limit is being exceeded or that the output device is shorted.

## Trouble-shooting

**Diagnostic Indicators.** The LED status indicators can provide much information regarding the field device, wiring, and the I/O module. Most input/output modules will have at least a single indicator. Input modules will normally have a *"power"* indicator, while output modules will normally have a *"logic"* indicator.

For an input module, the power LED ON indicates that the input device is activated and its signal is present at the module. This indicator alone cannot isolate malfunctions to the module, so some manufacturers provide an additional diagnostic indicator, the logic indicator. The logic LED ON indicates that the input signal has been recognized by the logic section of the input circuit. If the logic and power indicators do not match, the module is unable to transfer the incoming signal to the processor correctly.

The output module's logic indicator functions similarly to the input logic indicator. When it is ON, it indicates that the module's logic circuitry has recognized a command from the processor to turn ON. In addition to the logic indicator, some output modules will incorporate a blown fuse indicator, a power indicator, or both. The blown fuse indicator simply indicates the status of the protective fuse in the output circuit. When ON, the output power indicator shows that power is being applied to the load. Like the power and logic indicators of the input module, if both are not ON simultaneously, the output module is malfunctioning.

LED indicators provide much assistance in trouble-shooting; with both power and logic indicators, a malfunctioning module or circuit can be immediately pinpointed. LED indicators, however cannot diagnose all possible problems, but instead serve as preliminary signs of system malfunctions.

**Diagnosing Input Malfunctions.** In the case of an input malfunction, the first check is to see if the LED power indicator is responding to the field device (i.e. pushbutton, limit switch, etc.). If the input device is activated but the indicator does not energize, then the next test is to take a voltage measurement across the input terminal to check for the proper voltage level. If the voltage level is correct, then the input module should be replaced. If an LED logic indicator is illuminated and according to the programming device monitor the processor is not recognizing the input, then the input module may have a fault. If a replacement module does not eliminate the problem and wiring is assumed correct, then the I/O rack or communication cable should be suspected.

**Diagnosing Output Malfunctions.** In the case of output malfunctions, the first check is to see if the output device is responding to the LED status indicators. If an output rung is energized, the module indicator is ON, and the output device is not responding, then the field wiring should be suspected; but first, a check should be made for a blown fuse or the module simply replaced. If the fuse is OK and the replacement module does not solve the problem, then the field wiring should be checked. If according to the programming device monitor, an output device is being commanded to turn ON, but the indicator is OFF, then the module should be replaced.

When diagnosing input/output malfunctions, the best method is to isolate the problem either to the module itself or to the field wiring. If both power and logic indicators are available, then module failures become readily apparent. Normally, the first test is to replace the module or take a voltage measurement for the proper voltage level at the input or output terminal. If the proper voltage is measured at the input terminal and the module is not responding, then the module should be replaced. If the replacement module has no effect, then field wiring should be suspected. A proper voltage level at the output terminal, while the output device is OFF, indicates an error in the field wiring. If an output rung is activated and the LED indicator is OFF, then the module should be replaced.

If a malfunction cannot be traced to the I/O module, then the module connectors should be inspected for poor contact or misalignment. Finally, check for broken wires under connector terminals and cold solder joints on module terminals.

# CHAPTER
# --]11[--

# DATA HIGHWAYS

*Today's demands for higher productivity and quality products require that factory and machine control systems be integrated throughout the plant to achieve greater reliability and efficiency.*

As control systems become more complex, they require more effective communication schemes between the various system components. Some machine and process control systems require programmable controllers to be interlocked so that data can be passed among them to accomplish the control task efficiently. Other systems require a communication scheme that will allow centralized functions, such as data acquisition, system monitoring, and maintenance diagnostic and management production reporting, to be designed into a plant-wide system to provide maximum efficiency and productivity.

This chapter will present *data highways* and the role they play in achieving factory integration. It also includes a comparison of several communication networks that are currently in use.

## 11-1   GENERAL PRINCIPLES

### Definition

The term data highway is used in this book to describe an industrial *local area network*. The most general definition of a local area network (LAN) is "a high-speed medium-distance communication network." Data highways are used to communicate between several independent PC processors and/or host controllers that are located throughout a factory. Any device connected to the network or highway is referred to as a *node*. The maximum distance between two nodes on the network is usually given as at least one mile. Most definitions of a local network usually require that it supports at least 100 stations; however, there are networks that support as few as ten — or several thousand nodes. The transmission speed ranges from 56 kilobaud to 10 megabaud. An industrial network is one which meets the following criteria:

- Capable of supporting real-time control
- High data integrity (error detection)
- High noise immunity
- High reliability in harsh environment
- Suitable for large installations

Other common types of local networks are *business-system* networks, such as Ethernet, and *parallel-bus* networks, such as Cluster/One. Business networks do not require as much noise immunity as industrial networks because they are intended to be used in an office environment. The access time requirements are also less stringent. A user of a business work station can easily wait a few seconds for information, but a machine being controlled by a PC may need information within milliseconds. Parallel-bus networks are intended for microcomputers and minicomputers in office environments over short distances.

### Advantages

Before data highways came into use, two methods were used to communicate between PCs. The first method was to connect an output card of one PC to an input card of a second PC via a pair of wires. This method, which provided only one bit of information per pair of wires, could be expensive to install and very cumbersome to use. For a large installation, wiring costs can be greatly reduced through the use of a data highway. In the second method, PCs communicated through their programming ports via a central computer that was usually customer-supplied and programmed. This method limited the data throughput to the baud rate of the PCs programming port (as do some data highways) and has the disadvantage that if the central computer fails, the network becomes unusable (See Star Topology). Through the use of data highways, large amounts of usable data can be exchanged among PCs and any other hosts in a very efficient manner, through a dedicated communication link.

## Applications

The most common applications of data highways are *centralized data acquisition* and *distributed control*. Data collection and processing performed in individual controllers can burden the processors scan time, consume large amounts of memory, and tends to complicate the control logic program. These disadvantages can be eliminated by providing a data highway configuration in which all data is passed to a host computer that will perform any data processing. In distributed control applications, control functions once performed by a single controller are distributed among several controllers. Besides eliminating the disadvantage of dependence upon a single controller, performance and reliability are usually improved.

In order to allow centralized data processing and distributed control, the data highway and the PCs connected to it must provide the following functions:

- Communication among programmable controllers and other hosts
- Upload to a programmer or host computer from any PC
- Download from a programmer or host computer to any PC
- Read/Write I/O values/registers of any PC
- Monitoring of PC status and control of PC operation

## 11-2   TOPOLOGIES

The topology of a local area network defines how individual nodes are connected to the network. The major factors affected by topology are throughput, implementation cost, and reliability. The basic topologies presently used are: *Star, Common Bus,* and *Ring*. It should be noted, however, that a large network, such as the one shown in Fig. 11-1, may consist of a number of interconnected topologies.

## Star

As previously mentioned, the first PC networks consisted of a multi-port host computer with each of its ports connected to the programming port of a PC. This arrangement is known as a "star" topology and is shown in Fig. 11-2. The *network controller* shown can be either a computer, a PC, or other intelligent host. Most commercial computer installations are star networks in which many terminals are tied to a central computer. The main advantage of this topology is that it can be implemented with a simple point-to-point protocol. Each node can transmit whenever necessary. If error checking is not required, or a simple parity bit per character will suffice, then a "dumb" terminal can be a node. The star topology, however, has the following disadvantages:

- It does not lend itself to distributed processing due to dependence on a central node.
- The wiring costs are high for large installations.
- Messages between two nodes must pass through the central node, resulting in low throughput.
- Low throughput due to limited bandwidth.
- Failure of the central node will bring down the network.

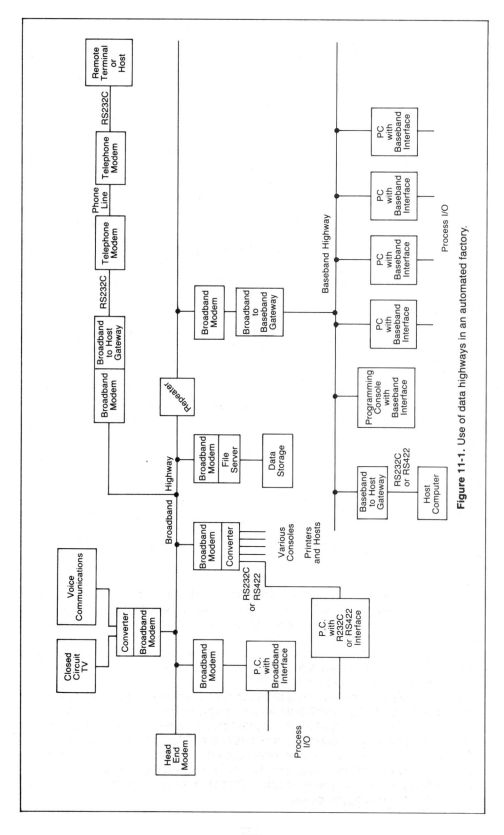

**Figure 11-1.** Use of data highways in an automated factory.

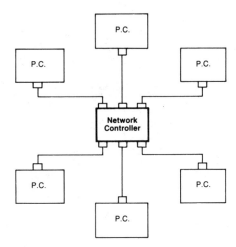

**Figure 11-2.** Star topology.

## Common Bus

The common bus topology is characterized by a main trunkline in which individual nodes are connected to a PC in a *multi-drop* fashion as shown in Fig. 11-3. Contrasted to the star topology, communication can take place between any two nodes without passing information through a network controller. An inherent problem, however is determining which node may transmit at any given time so that data collision is avoided. Several methods *(access methods)* of communication have been developed to solve this common bus problem.

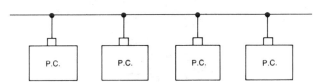

**Figure 11-3.** Bus topology with masterless access.

Common bus topologies are very applicable to distributed control applications since each station has equal independent control capability and can exchange information at any given time. In case a station must be added or removed from the network very little reconfiguration is needed. The bus topology generally uses a coaxial cable, with proper terminators, as the communication media. The main disadvantage is the dependence of a shared bus to service all the nodes. A break in the trunkline could affect many nodes.

Another implementation of the bus topology consists of several *slave controllers* and a network controller which acts as a *master* (see Fig. 11-4). In this configuration, the master sends data which is received by the slaves; if data is needed from a slave, the master will poll (address) the slave and wait for a response. No communication takes place unless it is first initiated by the master. The master/slave bus topology is usually implemented using two pairs of wires. On one pair of wires, the master sends, and all the slaves receive. On the other pair of wires, all the slaves transmit and the master receives.

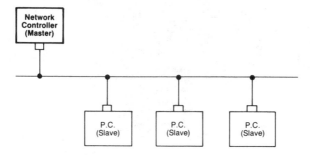

**Figure 11-4.** Bus topology with master/slave access.

## Ring

The "ring" topology, shown in Fig. 11-5, has not found its way into the industrial environment because failure of any node (not just the master) will bring down the network unless the failed node is bypassed. It is mentioned in this chapter, however, because it does not require multi-dropping and is a good candidate for a fiber optic network.

### Star-Shaped Ring

Some LAN manufacturers, like Proteon, have overcome the problem of node failure through the use of a *wire center*. The wire center, shown in Fig. 11-6, allows failed nodes to be automatically bypassed. As can be seen, however, this requires twice as much wire as the star topology. Therefore, it would have to offer some other significant advantage (such as use of fiber optics) before being applied to large installations.

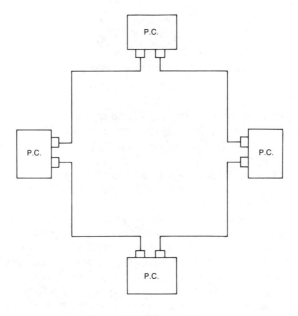

**Figure 11-5.** Ring topology.

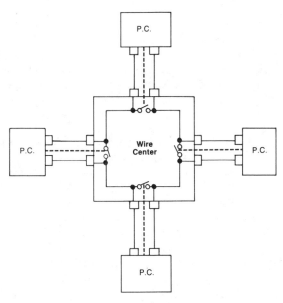

**Figure 11-6.** Star-shaped ring topology.

## 11-3   ACCESS METHODS

The access method is a manner in which the PC gains access of the highway for transmitting information. The bus type topology requires an access method since there are several controllers that must take turns to transmit data on the medium. This process requires, first of all, that the node has some means of disabling its transmitter in such a way that it does not interfere with the operation of the network.

This operation can be done in one of the following ways:

- Use of a modem which can turn off its carrier.
- Use of a transmitter which can be put into a high-independence state.
- Use of a passive current-loop transmitter which shorts when inactive and is wired in series with the other transmitters.

The most commonly used access methods are: *polling, collision detection, and token passing.*

### Polling

The access method most often used in master-slave bus configuration is known as "polling." Polling is a technique by which each station (slave) is polled, or interrogated, in sequence by the master, to see if it has data to transmit. The master sends a message to a specific slave and waits a specific amount of time for the slave to respond. The slave should respond by sending data or a short message signifying that it has no data to send. If the slave does not respond within the alloted time, the master assumes that the slave is dead and continues to poll the other slaves. Interslave communication in a master/slave configuration is inefficient since polling requires data to be first sent to the master and then to the receiving slave. Since polling is used in master/slave configurations, it is generally referred to as the *master/slave access method.*

239

## Collision Detection

Collision detection is also referred to as CSMA/CD for *Carrier Sense Multiple Access with Collision Detection*. In this method, each node that has a message to transmit waits until there is no traffic on the network and then transmits. While the node is transmitting, the collision-detection circuitry is checking for the presence of another transmitter. If a collision (two nodes transmitting at the same time) is detected, the transmitter is disabled and the node waits a variable amount of time before retrying. This method works well as long as there is not an excessive amount of traffic on the network. However, each collision and retry takes time which cannot be used for transmission of data. Therefore, the throughput drops off and access time increases. For this reason, collision detection has not been popular in control networks. It is, however, quite popular in business applications. Collision detection can be used for data gathering and program maintenance in large systems and in real-time distributed control applications with a relatively small number of nodes.

## Token Passing

Token passing is an access technique that is used to eliminate contention among PC stations that are trying to gain access to the highway. Token passing can be thought of as a form of distributed polling. A token is a message granting a polled station the exclusive but temporary right to control the highway (i.e. transmit information). This right, however, must be relinquished to a next designated node upon termination of transmission.

In a common bus network configuration using the token pass technique, each station is identified by an address. During operation, the token is passed from one station to the next in a sequential manner. The node which is transmitting the token also knows the address of the next station to receive the token. In token passing, one or more information packets containing source, destination, and control data circulate in the network. This information is received by each node and taken if needed. If the node has information to send it sends it in a new packet.

In a typical scenario, station 10 passes the token to station 15 (next address), which in turn will pass the token to station 18 (next address of 15). If a next station does not transmit to its successor, within a fixed amount of time *(token pass timeout),* then the token-passing station assumes that the receiving station has failed. In this case the originating station starts polling addresses until it finds a station that accepts the token. For instance, if 15 fails, station 10 will poll 16 and 17 without response since they are not present in the highway, and then polls 18, which responds to the token. The new receiving station will become the new successor and will remove the failed station from the network (18 will be 10's next address). The failed station will be patched-out.

The time required to pass the token entirely around the network is dependent on the number of nodes, and can be approximated by multiplying the token holding time by the number of nodes in the network. The token pass access method is preferred in applications requiring distributed control containing many nodes or having stringent response time requirements.

## 11-4   TRANSMISSION MEDIA

If installed properly, most data highways can be connected with any of the media discussed in this section. The installation includes the appropriate physical connectors and the correct electrical terminations. Media types commonly used for PC networks include *twisted-pair conductors, coaxial cables,* and *optical fibers.* The

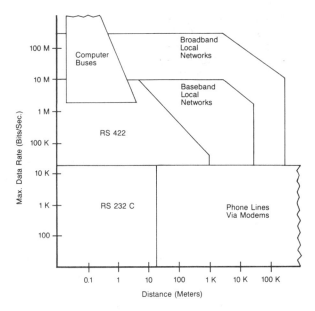

**Figure 11-7.** Communication comparisons.

type of media used and the number of nodes installed will affect the performance of the network (i.e. speed and distance). Figure 11-7 shows a comparison of different communication methods used with these media.

## Twisted Pair

Twisted pair has been used extensively in industry for point-to-point applications, over distances of up to 4000 feet and transmission rates as high as 250 kilobaud. Twisted-pair is relatively inexpensive and has fair noise immunity, which is improved if shielded. Performance drops off rapidly as nodes are added to a twisted pair bus. TIWAY I (Texas Instruments) and Modbus (Gould Modicon) both specify twisted pair buses with maximum transmission rates of 19.2 kilobaud.

## Baseband Coax

The performance limitations of twisted pair are mainly due to its nonuniformity. The characteristic impedance varies throughout the cable, making refflections difficult to reduce because there is no "right value" of termination resistor to use. Coaxial cable, however, is extremely uniform, thus eliminating reflections as a problem. The limiting factor then becomes capacitive and resistive loss. Baseband coax has been used in data highways at speeds up to 2 megabaud and distances up to 18,000 feet. It is usually 3/8 inch in diameter and costs about 20 cents per foot.

## Broadband Coax

Broadband coax is 1/2 to 1 inch in diameter and has been used for years to carry cable television signals. It can support a transmission rate of 150 megabaud and costs 30 to 50 cents per foot. Although this type of coax can be used to increase distance on a baseband network, it is intended to be used on a broadband network. These networks use frequency division multiplexing to provide many simultaneous channels. Each channel has a different RF carrier frequency. Broadband data highways use just one of these channels and one of the access methods previously mentioned. The transmission rate on the channel is typically 1, 5 or 10 megabaud. Broadband networks are capable of covering as much as 30 miles through the use of bidirectional repeaters and can support thousands of nodes. One advantage of using broadband is that the data highway can be implemented on an existing broadband network. The other channels of such a network are typically used for video, computer access, and various monitor and control functions.

Each broadband channel consists of two channels — a high-frequency forward channel and a low-frequency return channel. If only two nodes need to communicate, one can transmit on the forward channel and the other can transmit on the return channel. In a multidrop network, a head-end modem is needed to retransmit the return channel on its corresponding forward channel. The repeaters amplify the forward channel signals in one direction and return channel signals in the other direction. A broadband network with a baseband subnetwork was shown in Fig. 11-1.

## Fiber Optics

The main shortcoming of fiber optics today is that a low-loss tap has yet to be developed. This deficiency eliminates fiber optics for use in a large bus topology, but not from the star or ring topologies. In addition, fiber optics is three to four times more expensive than baseband coax, and optical couplers are several times more expensive than strictly electrical interfaces. Fiber optics does, however, have some impressive advantages. First, it is totally immune to all kinds of electrical interference. Second, it is small and light-weight. Transmission rates of up to 800 megabaud have been achieved at distances of 30,000 feet. In light of these facts, it would appear that fiber optics is bound to increase in industrial usage as the technology develops.

## 11-5    INTERPRETING THE SPECIFICATIONS

The purpose of this section is to assist the reader in determining if a particular data highway being considered can support a given application. Table 11-1 has been provided for the purpose of comparing the data highways offered by various vendors. It lists all of the readily available specifications for each highway, but some specifications, such as response time, can be difficult to determine. Each aspect of the data highway which must be considered is examined in the following topics.

### Maximum Number of Devices

For the system being considered, the system designer must determine how many nodes are required on the network and what type of device is to be used at each node. The device may be a PC, a vendor-supplied programmer, a host computer, or an intelligent terminal. As discussed, it must be determined if the highway will support each type of device and exactly how that device will be interfaced to the highway (hardware and software). If the device is a PC, the model of the PC must also be chosen. It should be noted that some models may not have the capability for being interfaced to the data highway. The data highway must be capable of supporting the number of nodes required for the current application, plus some reasonable number of nodes for future expansion.

### Maximum Length

The length of a data highway is generally specified in two parts: the maximum length of the main cable and the maximum length of each drop cable which is used between the device and the main cable. The values given in Table 11-1 are for main trunkline. The drop lengths are usually in the range of 30 to 100 feet. It is recommended, however, that drop lengths be kept as short as possible, since any drop will introduce some reflection onto the highway. The ideal case (electrically) is to run the main cable right to the device and back out again, even though this procedure increases wiring costs somewhat. Another important piece of information that must be obtained from the vendor is the type of cable which has to be used to achieve the specified length. For instance, the maximum lengths specified in Table 11-1 are dependent on the type of medium used. If the system needs the maximum length, the proper type of cable must be used. If the system requires a much shorter cable, money can usually be saved by using a less expensive cable.

## Response Time

Response time, as used in this book, means the time between an input transition at one node and the corresponding output transition at another node. As shown in the following equation, response time is the sum of the time required to perform the operations necessary to detect the input transition, transmit the information to the output node, and operate the output.

$$RT = IT + 2 \times ST1 + PT1 + AT + TT + PT2 + 2 \times ST2 + OT$$

where:

$IT$ = input delay time: the electrical delay involved in detecting the input transition

$ST1$ = scan time for sending node

$ST2$ = scan time for receiving node

$PT1$ = processing time: for the sending node, between solving the program logic and becoming ready to transmit the data

$PT2$ = processing time: for the receiving node, between receiving the data and having data ready to be operated on by the program logic

$AT$ = access time: time involved in becoming ready to transmit and transmitting

$TT$ = transmission time: time to transmit the data (this is the only time which is directly proportional to baud rate)

$OT$ = output delay time: the electrical delay involved in creating the output transition

The scan time includes I/O update time and any other overhead time, as well as program logic execution time, and can be easily defined as the time between I/O updates. The scan time is doubled in the above equation to include the case where the input signal changes just after the I/O update. In this case, the logic is first executed with the old information, the I/O update is done, then the logic is executed with the new information. This causes a two-scan delay.

I/O delay times and scan times are readily available values. Transmission time can be determined once the data rate and frame length are known. The data rate can be equal to the baud rate, but it is usually less. Synchronous systems have a data rate which is one half of the baud rate. Asynchronous systems have a data rate which is 80% of the baud rate due to the start and stop bits which accompany each 8 data bits. The access time and the two processing times are dependent upon the particular installation and generally have to be obtained from the manufacturer. If equipment is available, it is much easier and more accurate to deter-

**Table 11-1.** Sample highway specifications.

| Manufacturer | Highway Name | Max. Nodes | Max. Length (ft.) Max. Baud Rate | Baud Rate (max) | Access Method | Comments |
|---|---|---|---|---|---|---|
| Allen-Bradley | Data Highway | 64 | 10,000 | 56 K | Token | |
| GTE Sylvania | Control Net | 254 | 5,000 | 1M | Token | X.25 Compatible Gateway |
| General Electric | GEnet | 999 | 15,000 | 5M | Collision Detection | IEEE 802.3 Compatible |
| Gould-Modicon | Modbus | 247 | 15,000 | 19.2K | Master-Slave | Uses 2 Twisted Pairs, Shielded |
| Gould-Modicon | Modway | 250 | 15,000 | 1.544M | Token | |
| Industrial Solid State Controls | Copnet | 254 | 32,000 | 115.2K | Master-Slave | Includes interfaces to Allen-Bradley, Gould-Modicon and Texas Instruments PC's. HDLC Protocol. |
| Measurex | Data-Freeway | 63 | 10,000 | 1M | Collision Detection | |
| Reliance Electric | R-Net | 255 | 12,000 | 800K | Token | ASCII/X3.28/HDLC Gateway Uses HDLC Framing |
| Square D | SY/Net | 200 | 2000 | 500K | Timed Token | Worst case access = 600 msec with 50 PC's. |
| Texas Instruments | TIWAY I | 254 | 10,000 | 115.2K | Master-Slave | HDLC Protocol |
| Texas Instruments | TIWAY II | $2^{32}$ | 32,000 | 5M | Token | Broadband IEEE 802.4 compatible |
| Westinghouse | WDPF | 254 | 18,000 | 2M | Token | 100 msec Fixed Access Time. 10,000 Points/Sec Throughput |
| Westinghouse | Westnet | 50 | 10,000 | 1M | Master-Salve | Gateway Interface to Westnet |
| Comments | | | Broadband networks cover unlimited distances using CATV repeaters. | Lower baud rates allow greater lengths | | |

mine the overall response time through actual measurements. A simple procedure for performing this measurement is presented in section 11-7.

It should be noted that the parameter which needs to be determined here is not the average response time, but the maximum response time. Therefore, steps must be taken to create a worst-case environment during response time measurements. Creating this scenario involves tasks such as downloading programs and monitoring points while taking the measurements; this sort of activity increases PC scan times and data highway access times.

## Throughput

Some manufacturers specify the data highway throughput. This value usually represents the number of I/O points, which can be updated per second through the data highway. The throughput value is not enough to derive actual values for access time and data rate, even though it gives the system designer some idea of these values. In addition, throughput is sure to vary with system loading as a result of each node's processing time. Therefore, to have an accurate measurement of throughput, the conditions under which the measurement was taken must be known.

## Devices Supported

When considering each device in the system, the designer must ask not only *"will the data highway support this device,"* but also *"what is involved in connecting the device to the data highway."* For user-supplied devices, the designer must also determine what support software will be required.

**Programmable Controllers.** Each of the available data highways support at least some of the manufacturer's PCs. The PC is generally connected to the data highway through a separately-purchased interface unit. The interface unit connects to the PC either through a high-speed parallel bus or through the PC's serial programming port. In the latter case, two additional terms must be added to the response time equation: the transmission time on the programming port and the programming port processing time.

**Programming Devices.** Although most manufacturers offer some type of programming device which can be connected to the data highway, some manufacturers do not. In this case, all programming must be done through the programming port of the individual PC. A programming unit connected to the data highway provides centralized programming of any PC on the highway and also  various monitoring and control functions if available.

**Hosts.** Host support usually means that a user-supplied host computer can be used to perform programming functions, if it is programmed to conform to the manufacturer's protocol. It is usually connected to the data highway through a device called a *gateway*. The gateway contains a data highway port and another port (usually RS-232), which is connected to the host. This gateway greatly simplifies the software which the user must write for the host because the host-to-gateway link requires only a simple point-to-point protocol, rather than the masterless multidrop protocol of the data highway. The gateway also provides the appropriate electrical interfaces. Since most computers can provide an RS-232 port, no additional hardware is usually required.

**Intelligent Terminals.** The type of intelligent terminal referred to here is actually a small host computer complete with operating system and mass storage. It is interfaced to the data highway in exactly the same way as a large host computer. Anyone considering using one of these terminals on a data highway should investigate the software requirements closely to determine if the terminal's operating system will support the requirements. Some operating systems, for instance, provide for the transmission of only ASCII data, not binary data.

**Gateways.** In addition to the host gateway previously mentioned, some manufacturers may provide gateways to other multidrop networks. The other network may be another data highway (same or different manufacturer). There are also other types of host gateways; for instance, the host interface might be a high-speed RS-422 synchronous type. In this case, a protocol designed for synchronous use, such as HDLC, would probably be used.

## Application Interface

The question to answer here is *"how does the application program being executed by each PC allow it to share information with the other PCs?"* Most manufacturers provide at least one of the following methods.

- Reading of registers in other PCs
- Writing to registers in other PCs
- Reading and writing of "global" points or registers

For example, the state of an input on one PC can be detected by another PC on the network through the use of a global coil and global contact as follows:

```
        PC#1                          PC#2
     1011   G001                   G001   0015
     --] [------( )--              --] [------( )--
```

Whenever the global coil (G001) in PC#1 is energized, the global contact on PC#2 will close. This contact can be used like any other contact in PC#2's ladder program. It is up to the user to insure that each global coil is used by only one PC on the network. Reading and writing of registers is usually done via functional blocks. Care must be taken to assure that the capabilities provided are sufficient to support the communications needs of the application.

## 11-6 PROTOCOLS

A *protocol* is a set of rules which must be followed if two or more devices are to communicate with each other. The protocol includes everything from the meaning of the data to the voltage levels on the wires. The protocols define how the following problems are to be handled:

- Communication line errors
- Flow control to keep buffers from overflowing
- Access by multiple devices
- Failure detection
- Data translation
- Interpretation of messages

### ISO

In 1979, the International Standards Organization published its Open Systems Interconnection Reference Model. This model divides the various functions that protocols must perform into seven hierarchical layers. Each layer needs only to interface with its adjacent layers and is unaware of the existence of the other layers. The layers are defined as follows:

Layer 7 - Application (user interface)
Layer 6 - Presentation (data conversion)
Layer 5 - Session (establishment and disconnection)
Layer 4 - Transport (end-to-end service)
Layer 3 - Network (routing)
Layer 2 - Data link (error detection, framing)
Layer 1 - Physical (electrical characteristics)

Strictly speaking, only layers 1 and 7 are absolutely required. The other layers need only to be added as more services are required (such as error-free delivery, routing, session control, data conversion, etc.). Most of today's data highways contain layers 1, 2, and 7. As networks and protocols develop, the other layers are being implemented as required to allow connection to other networks.

### IEEE 802

The IEEE Standards Project 802 was established by the IEEE Computer Society in 1980 for the purpose of developing a local network standard. Such a standard would allow equipment of different manufacturers to communicate via the local network. After studying all the users' requirements and all the manufacturers' desires, the committee decided to produce a standard that defines several types of local networks. Figure 11-8 shows the scope of the standard, which at the publishing of this book was in its final draft stage. Industrial users and PC manufacturers have been showing a great deal of interest in both the broadband and baseband versions of the token bus option.

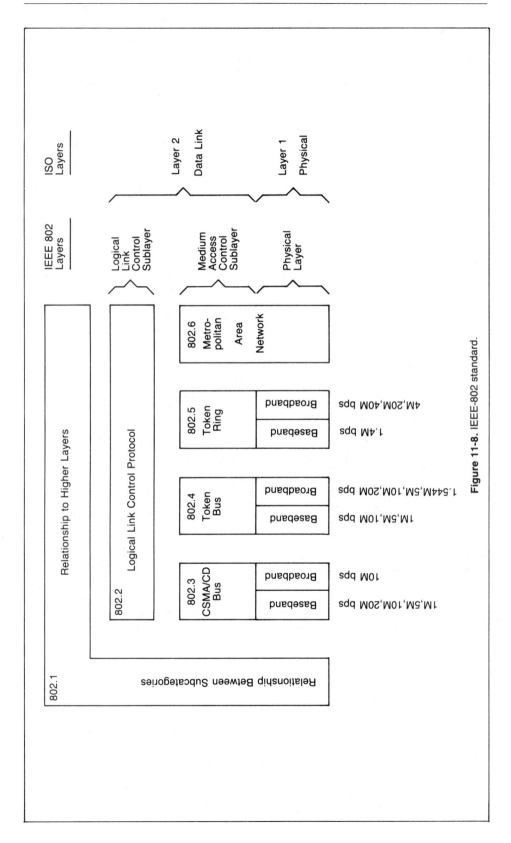

**Figure 11-8.** IEEE-802 standard.

## Proprietary

All the data highways presented in Table 11-1 use protocols which were devised by the manufacturers. None of these protocols has been proposed as a standard, but some of the specifications are available upon request to the manufacturer.

## ANSI

The American National Standards Institute has produced standard protocols for layers 2 and 3 that have been widely accepted. The most well-known and widely used of these is X.25, which specifies layer 2 and 3. It has been implemented on many host computers to interface to publicly switched networks. The X3.28 standard specifies layer 2 for point-to-point or multidrop service with a variety of options. This standard is used by many PC manufacturers on their programming ports.

## EIA

The Electronic Industries Association produced the RS-232C standard to define the physical interface (layer 1) between a computer or terminal (DTE) and a modem (DCE). This standard was rapidly adopted by modem manufacturers and has provided users with a wide selection of compatible modems. Although the standard did not address interconnection of two DTEs (such as a computer and a terminal), it was adopted for this use by crossing the transmit and receive wires in the cable. The maximum transmission rate is 19.2 kilobaud. Although the specified maximum distance is only 50 feet, it has been used for distances up to several hundred feet. The RS-422 standard defines a physical layer which can support transmission rates of up to 10 megabaud at distances of 200 to 4000 feet. Additional information on these standards can be found in section 3 of Chapter 6.

## 11-7   TESTING AND TROUBLE-SHOOTING

As mentioned in the previous section, before a data highway is installed, it should be tested to insure that it not only performs the desired function, but also provides the required response time. Application programming should be provided to monitor continuously the response time and take appropriate action if the response time goes beyond the maximum that the process will tolerate. This test may be performed by programming a *buzzer* circuit in which the contact closure is passed through every critical node on the network before it is returned. In this way, the pulse width of the created pulse is equal to the response time. This pulse width can be applied to a timer which is set to the maximum allowable response time. When the timer times out, the response time has been exceeded.

Trouble-shooting can be quite difficult unless both the manufacturer and the user take steps to simplify this task. The manufacturer can provide error counts and a self-test for each node. The user can provide application programming to detect the failure of each node. The extreme case of this would be to provide a "buzzer" and timer between each node and each other node. If the entire network goes down, it is probably due to a node with a transmitter which is shorted or constantly transmitting. The faulty node can be found by disconnecting each node, one at a time, and observing if communication is restored.

Some manufacturers provide a network monitor which can be used to detect a failed node, open cable, or excessive electrical interference.

# CHAPTER
# --]12[--

# INSIGHTS TO APPLICATIONS

*Application of programmable controllers and other automation technology has evolved from novelty to necessity.*

In this chapter, we will present an introduction to the most common areas in which PCs are being applied and a survey of their actual applications. Literally thousands of applications currently exist, and new ones are developing every day. To provide a detailed, useful explanation of each major application area would require a book in its own right. The intent here, however, is not to explain the applications, but to stimulate new ideas by showing examples of how PCs have been applied along with the benefits that were gained.

# 12-1   INTRODUCTION TO PC APPLICATIONS

In the early days, the programmable controller was an expensive electronic relay replacer. Although expensive, the PC was very beneficial to those who could afford the cost associated with equipment purchase and installation, as well as training of engineering and plant personnel. For this reason, the early PCs were more likely to be found in large U.S. manufacturing plants, such as Procter & Gamble, General Motors, and U.S. Steel. They were used to perform repetitive control functions that primarily involved simple timing, logic, and sequencing based on discrete inputs and outputs.

Application of today's controller has gone far beyond the simple control functions of its predecessors. Technology has not only made the controllers more capable, but also more affordable. PCs have become intelligent decision-making machines, with a wide scope of applications that now include variable control functions, data acquisition, report generation, and supervisory control. They also make possible a unique vision of inside the controlled machine or process via man/machine interfaces. Table 12-1 is just a sample list of how PCs are being applied. The potential benefits of PC application can no longer be simply neglected or reduced to a cost comparison with relays.

**Table 12-1. Examples of PC Applications.** This list of programmable controller applications — by no means an exhaustive list — was compiled by International Programmable Controls, Inc., Atlanta, Georgia.

---

### Rubber and Plastic

**Tire Curing Press Monitoring.** PCs perform individual press monitoring for time, pressure, and temperature during each press cycle. Information concerning machine status is stored in tables for later use and alerts the operator of any press malfunctions. Report generation printout for each shift includes a summary of good cures and press downtime due to malfunctions.

**Tire Manufacturing.** Programmable controllers can be used for tire press/cure systems to control the sequencing of events which must occur to transform the raw tire into a tire fit for the road. This control includes molding the tread pattern and curing the rubber to obtain the road-resistant characteristics. This PC application substantially reduces the space required and increases reliability of the system and quality of the product.

**Rubber Production.** Dedicated programmable controllers provide accurate scale control, mixer logic functions, and multiple formula operation of carbon black, oil, and pigment used in the production of rubber. The system maximizes utilization of machine tools during production schedules, tracks in-process inventories, and reduces time and personnel required to supervise the production activity and the manually produced shift-end reports.

**Plastic Injection Molding.** A PC system controls variables, such as temperature and pressure, which are used to optimize the injection molding process. The system provides closed-loop injection such that several velocity levels can be programmed to maintain consistent filling, reduce surface defects and stresses, and shorten cycle time. The system can also accumulate production data for future use.

### Chemical and Petrochemical

**Ammonia and Ethylene Processing.** Programmable controllers monitor and control large compressors that are used during the manufacturing of ammonia, ethylene, and other chemicals. The PC monitors bearing temperatures, operation of clearance pockets, compressor speed, power consumption, vibration, discharge temperatures, pressure, suction flow, and gas composition.

---

**Table 12-1. Examples of PC Applications (continued)**

**Dyes.** PCs monitor and control the dye processing used in the textile industry. They provide accurate processing of color blending and matching to predetermined values.

**Chemical Batching.** The PC controls the batching ratio of two or more materials in a continuous process. The system determines the rate of discharge of each material, in addition to providing inventory records on other useful data. Several batch recipes can be logged and retrieved automatically or on command from the operator.

**Fan Control.** PCs automatically control fans based on levels of toxic gases in a chemical production environment. This system provides effective measures of exhausting gases when a level of contamination is reached. The PC controls the fan start/stop, cycling, and speeds so that safety levels are maintained while energy consumption is minimized.

**Gas Transmission and Distribution.** Programmable controllers monitor and regulate pressures and flows of gas transmission and distribution systems. Data is gathered and measured in the field and transmitted to the PC system.

**Oil Fields.** PCs provide on-site gathering and processing of data pertinent to characteristics such as the depth and density of drilling rigs. The PC controls and monitors the total rig operation and alerts the operator of any possible malfunction.

**Pipeline Pump Station Control.** PCs control mainline and booster pumps for the distribution of crude oil. Measurement of flow, suction, discharge, and tank low and high limits are some of the functions they fulfill. Possible communications with SCADA (Supervisory Control And Data Acquisition) systems can    provide total supervision of pipeline.

## Power

**Plant Power System.** The programmable controller regulates the proper distribution of available electricity, gas, or steam. In addition, the PC monitors power house facilities, schedules distribution of energy, and generates distribution reports. The PC controls the loads during operation of the plant, as well as the automatic load shedding or restoring during power outages.

**Energy Management.** Through the reading of inside and outside temperatures, the PC controls heating and cooling units in a manufacturing plant. The PC system controls the loads, cycling them during predetermined cycles and keeping track of how long each load should be on during the cycle time. The system provides scheduled reports on the amount of energy used by the heating and cooling units.

**Coal Fluidization Process.** The controller can monitor how much energy is generated from a given amount of coal and regulate the coal crushing and mixing with crushed limestone. The PC monitors and controls burning rates, temperatures generated, sequencing of valves, and analog control of jet valves.

**Compressor Efficiency Control.** PCs are used to control several compressors located at a typical compressor station. The system handles safety interlocks, start-up/shutdown sequences, and compressor cycling and keeps the compressors running at maximum efficiency using the non-linear curves of the compressors.

## Metals

**Steelmaking.** The PC controls and operates furnances and produces the metal in accordance with preset specifications. The controller also calculates oxygen requirements, alloy additions, and power requirements.

**Loading and Unloading of Alloys.** Through accurate weighing and loading sequences, the system controls and monitors the quantity of coal, iron ore, and limestone to be melted. The unloading sequence of the steel to a torpedo car can also controlled.

**Continuous Casting.** PCs direct the molten steel transport ladle to the continuous-casting machine, where the steel is poured into a water-cooled mold for solidification.

**Cold Rolling.** PCs are used to control the conversion of semi-finished products into finished goods through cold rolling machines. The system controls the speed of the motors to obtain correct tension and provide adequate gauging of the rolling material.

**Aluminum Making.** Controllers monitor the refining process in which impurities are removed from bauxite by heat and chemicals. The system can ground and mix the ore with chemicals and then pump them into pressure containers, where they are heated, filtered, and combined with more chemicals.

**Table 12-1. Examples of PC Applications (continued)**

## Pulp and Paper

**Pulp Batch Blending.** The PC controls sequence operation, quantity measurement of ingredients, and storage of recipes for the blending process. The system allows operators to modify batch entries of each quantity if necessary and provides hard copy print-outs for inventory control and for accounting of ingredients used.

**Batch Preparation for Paper Making Process.** Applications include control of the complete stock preparation system for paper manufacturing. Recipes for each batch tank are selected and adjusted via operator entries. The system can also control the feedback logic for chemical addition based on tank level measurement signals. At the completion of each shift, the PC system provides management reports for material usage.

**Paper Mill Digester.** PC systems provide complete control of pulp digesters for the process of making pulp from wood chips. The system calculates and controls the amount of chips based on density and the digester volume; the percent of cooking liquors is calculated, and the required amounts are added in sequence. The PC ramps and holds the cooking temperature until the cook is completed. All data concerning the process is then transmitted to the PC for reporting.

**Paper Mill Production.** The controller regulates the average basis weight and moisture variable for paper grade. The system manipulates the steam flow valves, adjusts the stock valves to regulate weight, and monitors and controls total flow.

## Glass Processing

**Annealing Lehr Control.** PCs control the Lehr machine used to remove the internal stress from glass products. The system controls the operation by following the annealing temperature curve during the reheating, annealing, straining, and rapid cooling processes through different heat and cool zones. Improvements are made in the ratio of good glass to scrap, reduction in labor cost, and energy utilization.

**Glass Batching.** PCs control the batch weighing system according to stored glass formulas. The system also controls the electromagnetic feeders for infeed to and outfeed from the weigh hoppers, manual shut-off gates, and other equipment.

**Cullet Weighing.** PCs direct the cullet system by controlling the vibratory cullet feeder, weight-belt scale, and shuttle conveyor. All sequences of operation and inventory of quantities weighed are kept by the PC for future use.

**Batch Transport.** PCs control the batch transport system including reversible belt conveyors, transfer conveyors to cullet house, holding hoppers, shuttle conveyors, and magnetic separators. The controller takes action after discharge from the mixer and transfers the mixed batch to the furnace shuttle, where it is discharged to the full length of the furnace feed hopper.

## Materials Handling

**Storage and Retrieval Systems.** A PC is used to load parts and carry them in totes in the storage and retrieval system. The controller keeps tracking information such as storage lane number where parts are stored, the parts assigned to specific lanes, and quantity of parts in any particular lane. This PC arrangement allows rapid changes of parts loaded or unloaded from the system. The controller also provides inventory printouts and informs the operator of any malfunctions.

**Automatic Plating Line.** The PC controls a set pattern for the automated hoist which can traverse left, right, up, and down through the various plating solutions. The system knows at all times where the hoist is located.

**Conveyor Systems.** The PC controls all the sequential operations, alarms, and safety logic necessary to load and circulate parts on a main line conveyor as well as sorting product to the correct lane. The PC can also schedule lane sorting to optimize palletizer duty. Records are kept on a shift basis for production of good parts and part rejections if required.

**Automated Warehousing.** The PC controls and optimizes the movement of stacking cranes and provides high turnaround of materials requests in an automated high cube vertical warehouse. The PC also controls aisle conveyors and case palletizers to significantly reduce manpower requirements. Inventory control figures are maintained and can be provided on request.

## Automotive

**Internal Combustion Engine Monitoring.** The system acquires data recorded from sensors located at the internal combustion engine. Measurements taken include water temperature, oil temperature, RPMs, torque, exhaust temperature, oil pressure, manifold pressure, and timing.

**Table 12-1. Examples of PC Applications (continued)**

**Carburetor Production Testing.** PCs provide on-line analysis of automotive carburetors in a production assembly line. The systems significantly reduce the test time, while providing greater yield and better quality carburetors. Some of the variables that are tested are pressure, vacuum, and flow.

**Monitoring Automotive Production Machines.** The system monitors total parts, rejected parts, parts produced, machine cycle time, and machine efficiency. All statistical data are available to the operator and also at the end of each shift.

**Power Steering Valve Assembly and Test.** The PC system controls a machine to insure proper balance of the valves and maximize left and right turning ratios.

### Manufacturing/Machining

**Production Machines.** The PC controls and monitors automatic production machines at high efficiency rates. The piece-count production and machine status are also monitored. Corrective action can be taken immediately if a failure is detected by the PC.

**Transfer Line Machines.** PCs monitor and control all transfer line machining station operations and the interlocking between each station. The system receives input from the operator to check the operating conditions of line-mounted controls and reports any malfunctions. This arrangement provides greater machine efficiency, higher quality products, and lower scrap levels.

**Wire Machine.** The controller monitors the time and ON/OFF cycles of a wire drawing machine. The system provides ramping control and synchronization of electric motor drives. All cycles are recorded and reported on demand to obtain the machine's efficiency as calculated by the PC.

**Tool Changing.** The PC controls a synchronous metal cutting machine with several tool groups. The system keeps track of when each tool should be replaced, based on the number of parts it manufactures. It also displays the count and replacements of all the tool groups.

**Paint Spraying.** PCs control the painting sequences in auto manufacturing. Style and color information is entered by the operator or a host computer and is tracked through the conveyor until the part reaches the spray booth. The controller decodes the part information and then controls the spray guns to paint the automotive part. The spray gun movement is optimized to conserve paint and to increase part throughput.

## 12-2  A SURVEY OF PC APPLICATIONS

The following examples of programmable controllers in real-world applications are designed to give more than a theoretical view of how PCs are used. These examples describe the applicability of PCs to machine and process control, data acquisition and reporting, and interfacing man with machine or process. In addition to innumerable system benefits (e.g. efficiency, minimal downtime, reduction of personnel, higher quality products,  lower scrap levels, and higher throughput), PC installations almost always provide fast investment payback and additional savings.

**Application #1:** AUTOMATED CAR WASH (Allen-Bradley Co.)

**Controller:** PLC-2/20

**Why PC Was Applied:** The use of a programmable controller at a car wash was chosen for flexibility. As in many cases, the PC was selected for its easy reprogramming ability. Because new car-washing equipment is constantly being improved, the user wanted to be able to adapt the control systems easily to their new layout. The PC offers this flexibility. The control program is written so that a newly built car wash could be controlled using the same program, but with minor changes.

**Description:** When a car enters the building, it darkens a photoeye that measures the car's length. The photoeye feeds this information to the programmable controller, where it will modify each of the wash and dry cycles to match any car's length — from Escorts to Cadillacs. The attendant uses the operator's panel to tell the PC whether a car will receive a standard wash, a special wash, or wax, among other options. The selected option is displayed on a lighted panel above the first wash station.

The movement of the cars through the wash is measured with pulses sent from the conveyor chain that pushes the car through the wash. Each foot the car travels sends one pulse. A TTL input module accepts the pulse inputs, which are accumulated by an internal counter. Accumulated and preset values initiate each step in the wash and dry process. The length of the car wash is 180 feet, so the total number of pulses is 180. Ten pulses are added as a "time out" after the last car in queue leaves the wash.

The entire series of washing and drying operations was broken down into 24 steps. Each step is initiated by the value of the counter. When the counter has accumulated 140 pulses, for example, three air dryers are turned on. When the counter's accumulated value is 161, the car has passed the drying step, and the dryers turn off. The process continues similarly for each operation.

A set of 7 thumbwheels on the control panel allows the PC programmer to adjust the system, if required. The first two BCD-coded thumbwheels allow a two-digit input value to tell the PC which of the 24 total steps is to be modified. The next three thumbwheels allow a three-digit input value to tell the PC how many pulses should be counted before initiating the next operation. The last two thumbwheels allow extra pulses to be added to extend the operating length.

The program for this application required 7K of the PLC-2/20's 8K maximum memory. This arrangement was necessary because the series of operations must be repeated up to 10 times for each of the 10 cars the wash can handle simultaneously. The PC is also programmed to shut down the whole operation if 30 seconds pass without any cars in the wash.

The user also depends on the PLC-2/20 Programmable Controller to direct other components of the car wash, including three air dryers, a 500-lbs. water pressure rinse, and the retractable top brush for cars with luggage racks and antennas.

**Results:** The management of the car wash reported that energy, water, and maintenance costs have all been substantially reduced from this PC application.

### Application #2: PLATING LINE (Square D Co.)

### Controller: SY/MAX 300

**Why PC Was Applied:** Production from two metal caster plating lines was once plagued by persistent malfunctions and the necessity for frequent maintenance of the aging electromechanical control systems. A decision by the company to replace the faltering controls led to the installation of two programmable control systems. The new control systems were designed to incorporate operating features of the old systems with the technology and reliability of PCs.

**Description:** Each plating line involved thirteen tanks. Tanks numbered 5 through 9 are used for the aluminum and copper plating, and the remaining eight tanks are used for various rinses. With a selector switch as a PC input, the operator can vary the number of plating tanks that will be used on a particular day. To simplify the operator transition to the PC control systems, the design includes external timers mounted on the front of the enclosures, thus providing the same operator input features as the old system. With these timers, the operator can vary the dripping and rinsing times at the various tanks, which continually change to compensate for different base metals and external tank temperatures.

The line operators can monitor the position of the hoist at any stage along the plating lines by indicators mounted on each control panel. These indicators advise the operator as to which of the 47 total possible steps the hoist is on. A manual override capability incorporated in the system allows the operator to check individual functions of the lines and the capability to shutdown a line in the event of an emergency. A block-stop feature, designed in the systems, allows one additional step to occur on a plating line before halting the sequence. When the block stop is used, the operator must again depress the start button to begin another step.

**Results:** The result has been a substantial improvement in production. The previously used relay and cross-bar storage systems, which received information from motor driven tape readers, were replaced.

**Application #3:** CONTROL OF POWER AND MATERIAL FLOW
(Allen-Bradley Co.)

**Controller:** PLC 1774

**Why PC Was Applied:** The controller was applied to monitor and control the distribution of electric power and materials, required to build a dam. The application required an elaborate conveyor system to deliver to the embarkment 22 million cubic yards of material, situated in a borrow area 800 to 1200 feet above the dam's foundation.

Elaborate conveyor system moved 8500 tons of material to the dam site every hour — 20 hours a day (Courtesy Allen-Bradley Co.).

**Description:** The controller was programmed to be a power conservationist. It monitors power output and potential malfunctions, as well as flow and speed of materials. As overseer of the project's electrical requirements, the PC carefully modulates the total power output of the powerhouse, as well as the downhill belt. The powerhouse is more expensive to operate and is only called upon during peak load demands.

At low demand periods, the PLC 1774 sets the diesel generators at idle speed to conserve fuel. It will always favor the cheaper power source generated by the downhil belt. At very low demand periods, when the belt is producing an excessive amount of power, a computer routes electricity into a series of incremental resistor banks. In these banks power is consumed until additional loads brings it back on-line.

The control program also included instructions to fuse the total power output in the event of a rapid load shed, which could introduce a reverse power situation in the powerhouse. This situation is automatically prevented by causing an instantaneous blackout of the entire job, and at the same time, stopping the belt by applying the mechanical brakes.

Belt speed and weight of material are monitored by sensing devices on the main conveyor. Similar devices are mounted in a 1,150 ton surge bin and on the various feeders. These sensors constantly report how much material is on the way and how much room is available in the main bin. This information is transmitted every 12 seconds. The purpose of monitoring is to maintain a constant materials flow below the maximum capacity of the surge bin. This procedure will minimize the need to stop the downhill belt.

The PLC 1774 was programmed to satisfy the material requirement in the main hopper. It will automatically modulate the activities of all feeders, conveyors, and hoppers above and below it. This modulation will ensure a constant flow of material to the portable load-out bins delivering fill to the awaiting haulers.

**Results:** The entire system moved 8,500 tons of material to the dam site every hour for 20 hours each day. When the dam was topped out in 1981, the system was dismantled and made available for other projects or resale.

**Application #4:** FEED MIXING (Westinghouse Co.)

**Controller:** Numa-Logic PC-700

**Why PC Was Applied:** A feed company found it nearly impossible to achieve productivity and efficiency before applying a programmable controller. Prior to installing automatic controls, the feed was mixed manually in the mill by men using shovels, wheelbarrows, and plenty of muscle, in an extremely cold environment during the winter and an unbearably warm environment during the summer, within the confines of the year-round-operated mill.

The Numa-Logic PC-700 intallation during start-up (Courtesy Westinghouse Electric Corp.).

**Description:** Through the capabilities of the PC-700, the feed company was able to automate completely the manual batch mixing feed mill. The PC is responsible for selecting the proper amounts of the necessary ingredients for the feed formula desired. It also decides whether the selected mixture should be processed on a continous basis until the bin becomes filled to capacity or processed only in the amounts needed at a given time.

The batch mixing system is controlled by three PC-700 programmable controllers. The PC-700s have approximately 60 discrete inputs, 215 discrete outputs, 5 register outputs (to LED displays), 8 register inputs (thumbwheels), and 41 digital-to-analog (D/A) interface modules which control the variable speed DC drives. The PC controls a DC drive system on each of the 41 ingredient feeders, the conveyor system, the mixer, the blender, the scale, and the distribution system which distributes the properly mixed feed to the finished feed bins. Two remote input/output (I/O) stations are controlled by the three PC-700s as well. The total sytem distributed over three control panels, comprises a total of 54 I/O racks.

In addition to controlling production levels, the PC controls formula selections and the dust control system, which is very important due to the highly inflammable nature of the dust expelled when the finished feed formula is dumped into the bins. The PCs also control all emergency shutdowns and provide for special functions. These special functions offer the user the flexibility to perform such actions as clearing low ingredient conditions, calibrating ingredient flow, setting up special formulas, and moving raw ingredients to bins in the old mill.

**Results:** Automation of the feed company has brought increased production, consistent feed mixture, accurate feed measurement, safer and more comfortable working conditions, and greater efficiency. Having achieved a 300% increase in productivity, the feed company can create 30 different formulations of finished feed containing 41 ingredients with only a one percent weight error. The feed production rate went from 30 tons per hour to 90 tons per hour, for a total of 700 tons per day. In addition, maintenance of the automated system is virtually non-existent. This is a significant improvement to the almost daily maintenance required with the manual batch feeding system, a system which had outlived itself by 20 years.

**Application #5:** ENERGY MANAGEMENT AT THE SILVERDOME
(Gould Modicon, Inc.)

**Controller:** Model 484

**Why PC Was Applied:** Three programmable controllers serve as the basic control devices for the energy management system at the Silverdome. The controllers monitor and control energy usage by controlling the on and off states of fans, blowers, heaters, and other mechanical devices necessary for the most efficient operation of the all-season, 80,000-seat municipal stadium complex.

**Description:** The control system utilizes a customized distributed processing network that interfaces to three 484 controllers and a minicomputer via RS232 interfaces. The PCs monitor the heating, ventilating, and air conditioning devices, as well as the fire pull stations and security call stations. These PCs use 320 digital inputs and 160 digital outputs. The minicomputer acts as the network communications manager and provides an interface to the multiple operator terminals. Simultaneously, the minicomputer system can provide auxiliary functions for word processing, accounting, and high-level programming languages.

In the event of minicomputer failure, the PCs can provide stand-alone, fail safe operations. If necessary, each of the three controllers can operate independently of the others to maintain operations. The program interlocks time scheduling devices, optimizes start-stop scheduling, controls damper volume, sets back temperatures, and stages preheat/reheat and refrigeration, as well as stadium temperature and pressure control to maintain the air-supported domed roof. The control program manages the operation of 24 constant speed building pressure blowers and a single variable speed building pressure blower, as well as 10 concourse heating fan systems. The controller also selects the proper fans and/or heaters required to maintain the building setpoints during occupancies in either the summer or winter.

**Results:** When compared to the manually operated system, the programmable control system provided over 25% savings in energy consumption.

**Application #6: D**OUBLE-END TRIM AND BORE MACHINE
(Texas Instruments Corp.)

End Drill section of trim and bore machine (top), and double end trim machine with controller panel (Courtesy of Texas Instruments Corp.).

**Controller:** TI-510

**Why PC Was Applied:** The original double-end trim and bore machines were pneumatically controlled with air logic, using rocker valves to divert mechanically air flow and control air cylinder operation. This control was sluggish and unreliable in the dusty environment characteristic of woodworking shops. The pneumatic systems were eventually replaced by electromechanical relays. Although reliability was improved, the relay system was expensive to install and offered almost no flexibility.

**Description:** The machine operation is simple. A wood piece is inserted into the machine and the operator depresses the cycle start switch. The piece is clamped into place using an air cylinder to prevent it from slipping. With the piece in place, the machine then trims either or both ends of the piece by lowering two circular saws into the wood with a hydraulic cylinder. After the saws return to their upper position, the two end drills bore either or both ends to preset depth. The piece is then released and removed.

**Results:** The upgrading of the trim and bore machine has resulted in almost no downtime and now allows minor changes to be made easily.

**Application #7:** CONTROL OF DRUM-MIX ASPHALT PLANT
(Gould Modicon, Inc.)

**Controller:** Model 584

**Why PC Was Applied:** The controller was needed to replace the old generation of control systems for drum-mix plants with electronic control systems that would produce hot mix far more economically than the conventional plant.

**Description:** The 584 PC provides automatic sequential startup, shutdown, mix change capabilities, and plant safety interlocks, as well as continous control of aggregate and asphalt blending. Under the control of the 584, each drum-mix plant is capable of producing hot mix for 20 or more variable mix types from recipes stored in memory. The recipes can also be entered and changed in the manual mode.

For each mix, the required formula or percentage of each aggregate and asphalt is entered into the controller by digital inputs on the console. The desired production rate of hot mix asphalt is entered and the PC calculates the speed at which the feeder belt must operate to deliver the required amount of each aggregate.

If the mix is one of the pre-set formulas, the operator dials up the formula number and does not have to enter the recipe. The operator sets the weight of the asphalt in pounds per gallon at 60 degrees Fahrenheit and also sets the moisture percentage of the aggregate entering the drum mixer. After the information has been read by the controller, the process begins when the operator pushes the sequential start button.

To change mix, the operator dials another formula, sets the desired rate of production, and pushes the mix change button. Prior to starting the process with the new mix, the controller makes necessary calculations to adjust the feeder belt speeds and asphalt flow rate. This feature is instrumental in reducing the quantity of out-of-spec material to a minimum.

Data concerning plant operation can be continually gathered and printed-out in report form on demand. Useful reports such as up-to-the-minute printouts on mix produced, fuel usage, total tons produced, production rate, and new aggregate inventory by size can serve as an important management tool.

**Results:** With the programmable control system, drum-mix operations cost less to heat and require fewer personnel to operate the plant's, which reduces electrical costs as well as the cost of labor. Programmable controller capabilities enhanced the plants ability to track cost and to define the most efficient operating rates.

**Application #8:** SEMI-AUTOMATIC DUAL PALLETIZING SYSTEM
(Allen-Bradley Co.)

**Controller:** AB-MINI-PLC-2

**Why PC Was Applied:** The PC system was needed to reduce labor costs in building pallets and to keep an orderly flow of material through the plant. A decision to replace the manual system with an automatic one would also help to ensure maximum output.

Hampers entering pelletizers as directed by a diverter bar (Courtesy of Allen-Bradley Co.).

**Description:** The finished cookware is unloaded and placed on a monorail system which conveys the cookware to the final inspection area. In this area, seven accumulating conveyors empty onto a common trunk line conveyor. Each of the seven accumulating lines has two work stations where the product is hand examined for defects. Acceptable cookware is placed into large hampers which are about the size of orange crates. The inspector then places the hampers of inspected ware on a metering belt, which feeds onto an incline conveyor and transports the hamper 10 feet up to the overhead conveyor system. The entire conveying system is suspended overhead, allowing use of the floor area below.

A queue of hampers, 50 feet long, can gather on the seven accumulation conveyors. These hampers are lined up as they are fed from an inspection station and are prevented from entering the trunk line by an air-operated stop, located at the front of each accumulation conveyor. Each accumulation conveyor is allowed to accumulate a pallet load of twelve hampers, signaled by a photoeye input which is darkened when the last hamper is detected. Another stop is raised to prevent other hampers from entering the primary zone. If necessary, a secondary zone is available to handle six additional hampers in an overflow area.

When signaled by a darkened photoeye, the Mini-PLC-2 Processor checks the mainline conveyor for any hampers from the other six conveyors that are also presently accumulating. If the mainline is clear, then one queue of 12 hampers is discharged onto the mainline and is routed to one of the palletizers for processing. When all hampers released for a single palletizer have been counted by a photoeye at the palletizer, a diverter arm swings to allow the next load of 12 hampers to enter the opposite palletizer. The mainline is clear now, and the next accumulating conveyor can be emptied.

The two palletizers are designed to allow three levels, four hampers per level, to be stacked for a unit load. An operator stationed on a platform between the two palletizers uses a foot switch to control the flow of hampers. The operator orients four hampers in one operation on an indexing platen and then retracts and lowers them onto a pallet. This process is repeatd three times for each pallet, at which time a full pallet is detected by a tier counter, and then the pallet is lowered to the floor level where it is sent to final inspection. Upon reaching the final inspection station, the pallets are transported by fork lifts to the dock for shipment packaging and distribution.

In this application, a 64-I/O rack was used for each Mini-PLC controller with all AC input and output modules wired to photoeyes, limit switches, and other standard control devices. To allow production to continue, even if the automatic equipment goes down, multiple manual back-up systems are available to ensure maximum flexibility, including complete monitoring and manual override of the conveyors, trunk, and palletizers.

**Results:** Despite continuous 24-hour operation, there has been no significant downtime incidents related to failure of the automatic control system.

### Application #9: SOOT BLOWER CONTROL SYSTEM (GTE Sylvania)

**Controller:** MECA

**Why PC Was Applied:** The PC was needed to control and monitor 42 soot blowers.

**Description:** This system uses one MECA controller with 96 I/O. The soot blowers may be operated in sequence, automatically at the press of a button, or the operator may select specific soot blowers using a 2-digit thumbwheel switch.

The system is monitored at all times and its status displayed on a Cherry alphanumeric display. If a soot blower fails, the reason for the failure and location will be displayed. The PC also keeps track of which soot blowers have been successfully operated and which ones have failed. This information is indicated on a small graphic panel using a steady light to indicate a successful cycle, a slow flashing light to indicate a soot blower in operation, and a fast flashing light to indicate a soot blower that has failed.

**Results:** The new system offers significant improvements in flexibility over the old system, and it allows failures to be quickly located and diagnosed.

### Application #10: GRAIN ELEVATOR CONTROL (Gould Modicon, Inc.)

**Controller:** MODEL 484

**Why PC Was Applied:** The system was arranged to provide electronic control for new elevator systems and to update and modernize old elevators that have had few changes over the past 50 years.

**Description:** The electronic control system uses a Gould Modicon 484 programmable controller as the basic control device with an interface to a minicomputer. The 484 monitors and controls conveyors, slides, elevators, opening and closing of chutes, and grain weighing. The controller also monitors bin capacity, blends the grains, handles interlocks, and provides security and alarm signals as required.

A voice synthesizer that receives its signal from the 484 is used as an alarming and security device. On command from the 484, the synthesizer makes audible voice announcements in emergency situations, such as fire or break-in, or for a management alert of equipment breakdown. Five sequential phone alerts can be generated and directed to the fire department, police, plant manager, power company, or other persons on the emergency call list. The PC is also responsible for sending all emergency conditions to a printer for a hardcopy printout. For each emergency condition, the voice synthesizer repeats the message in the main control booth until it is manually turned off by the plant operator.

**Results:** Formerly, as many as five or six people were required to operate some of the co-op elevators with duties as operations manager, unloader, conveyor and chute operator, and other associated positions. With the electronic control system, an elevator can be operated on each shift with two or three operators, who manage and control the entire facility with ease.

**Application #11:** PIPELINE CONTROL (Reliance Electric, Toledo Scale Div.)

**Controller:** AUTOMATE 35

**Why PC Was Applied:** The programmable controller was applied to replace the old electromechanical control system and to simplify the tank-farm operation.

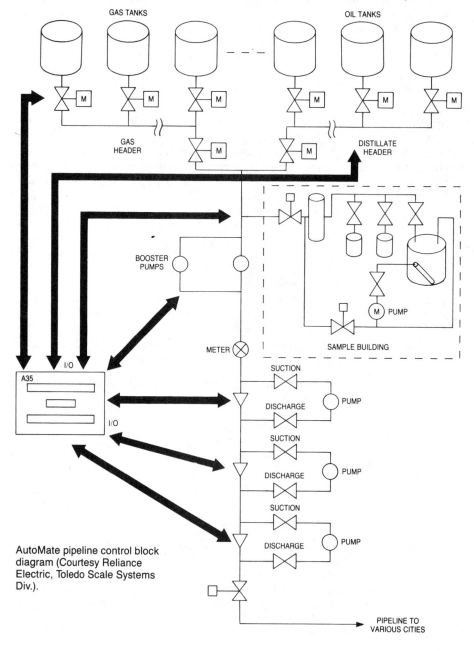

AutoMate pipeline control block diagram (Courtesy Reliance Electric, Toledo Scale Systems Div.).

**Description:** The pipeline control system hardware includes three AutoMate 35 processors and remote parallel output and input cards for pushbutton and operational interfacing. A CRT and cassette unit are used for programming and data storage. Each controller is mounted and wired in a separate cabinet that is located at the originating station control house.

Product selection is made by using local pushbuttons or remote interfacing signals. The products are divided into gas or oil and are located in storage tanks. The operator must select the next product which will be put into the pipeline, so that at the proper time a switchover can be made.

When a product change is initiated, the PC controls the changeover process. The control sequence consists of shutting down the current product, opening the correct valves for the new product, and starting and stopping appropriate motors and pumps. These actions must occur in the proper sequence. When the changeover sequence is complete, a header valve is selected and product flow is allowed.

As the product flows from the tank into the pipeline, it passes a sampling system. Under control of the AutoMate, the sampling system can extract samples of the product for quality analysis, if desired. Booster pumps are sequenced on and off to provide product flow to the units which pump the product to the pipeline. The PC also monitors sump levels and can monitor both case pressure and discharge pressure, if desired.

**Results:** The new control system allows easy changes to meet additional requirements and significantly reduces trouble-shooting time. Cost savings were realized by using remote input/output systems.

### Application #12: COLD STORAGE REFRIGERATION SYSTEM (GTE Sylvania)

**Controller:** MECA

**Why PC Was Applied:** The system was needed to provide efficient control of compressor cycles.

**Description:** This system uses two MECA programmable controllers-- one to control the air handlers and their associated gas valves and the other to control six compressors as well as to monitor all alarm functions.

The air handlers cycle on and off as determined by a thermostat. Periodically, each handler will go into an automatic defrost cycle. The operator may set the air handlers to defrost 1, 2, 3, or 4 times a day via a selector switch.

The PC cycles the compressors on and off as demand requires, adding compressors during high demand and dropping them during low demand. Each compressor, however, is only allowed to start twice per hour. If a certain compressor is called for and it has already been started twice in one hour, another compressor will be called for in its place. A high temperature alarm will override the two starts per hour rule, as well as interrupt any defrost cycles that may be in progress in the air handler system. A first-out annunciator system is also part of the MECA software.

**Results:** The MECA programmable control system resulted in energy savings by providing efficient control of compressor cycling, during high and low demand periods.

### Application #13: CEMENT PRODUCTION (Westinghouse Electric Corp.)

**Controller:** Numa-logic PC-700

**Why PC Was Applied:** The system was installed in an attempt to accomplish several improvements in the company's production of one million tons of cement a year. The main goals included implementing a plan to save in production and maintenance costs as well as time; to achieve better control over system operation; and to accomplish greater efficiency. The previous system's electromechanical relays, utilized at the silo sites, were hermetically sealed, but still required weekly replacement due to dust build-up. Such frequent replacement caused great expense in manpower, material, and production downtime.

**Description:** The basic method for producing cement begins with rock, such as limestone or shale. The rock is crushed and sent to a raw grind ball mill where it is pulverized into a fine powder, the consistency of talc. Next, the fine substance is sent to storage silos for blending. An elevator and screw conveying system air-pushes the the thoroughly-mixed powder along a material conveyor belt to a kiln, where it is baked at 2,700 degrees Fahrenheit and transformed into clinker, a material resembling small, black ball bearings. The clinker, once ground in the finish ball mill, is then known as cement and is ready to be loaded into rail cars, trucks, or bags. In this condition, it is ready for purchase by users, such as ready-mix cement plants, large construction firms, and other buyers. These users will then mix the cement with aggregate, sand, and water to make concrete.

The specific task to be met by the PC was to control the portion of cement processing from the raw grind ball mill to the silo and finally to the kiln. The PC-700 with remote I/O was used to control efficiently this intermediate process known as aeration. Aeration, an air fluffing process which uses air to blend the powder, keeps the fine mixture at an even consistency so that it burns uniformly in the kiln. The aeration process lasts a minimum of four hours and takes place in five silos, each of which has a 2,300-ton capacity.

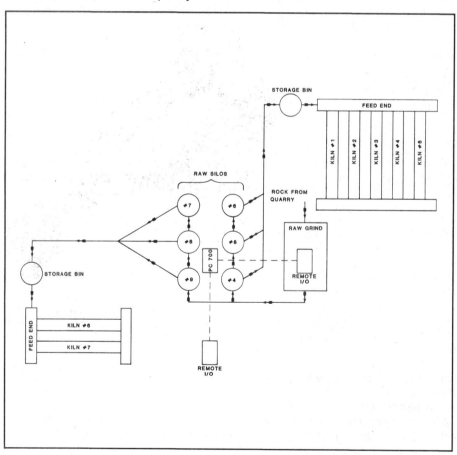

Block diagram of cement production (Courtesy of Westinghouse Electric Corp.).

Another important element of the aeration process is sequencing, which not only keeps the process running smoothly, but also most importantly, allows the controller to maintain air pressure and silo volume within safe limits, thus preventing explosion.

The PC installation includes two remote I/O locations. One remote I/O station is located in the raw grinder area and controls all raw silo functions. The second remote I/O station is located at the preheater kiln and monitors the height of the product in each of the raw silos.

**Results:** The system was installed at a cost of less than $15,000. Based on fuel savings alone, the system paid for itself in only four months. In fact, the company estimated that if even 1% in fuel reduction is realized yearly, an annual savings of $140,000 will result.

The new system also keeps the I/O and power supply in a clean environment, requires considerably less maintenance, and offers better operator control and greater efficiency.

**Application #14:** ENGINE TESTING (Allen-Bradley Co.)

**Controllers:** PLC 1774 & MINI-PLC-2

**Why PC Was Applied:** Engineers at a major automotive manufacturer chose the A-B controllers because of the data highway capability that allowed them to integrate their PCs without using computers, which would have placed additional burdens on their computer department. The PC utilized required a large memory area for data handling purposes.

**Description:** The total system consists of the PLC 1774 Controller, with 18 remote I/O racks and 12 Mini-PLC-2 controllers, which communicate over the data highway to the PLC Controller. The control system is responsible for putting diesel and gasoline engine alternators or starters through their paces.

Tests are conducted in two cells: one for starters and another for alternators. In the starter test cells, 18 gas or diesel engines, each connected to a remote I/O rack, are run through five different test cycles, controlled by one PLC processor.

The first test is a battery run-down cycle in which a starter must crank an engine without any ignition applied. The test continues until the battery voltage drops to 7.5 volts. This test ensures that the starter can endure the build up of heat from long cranking times.

Next is an automatic test in which the starter is subjected to 60,000 starts or until a failure occurs.

The third cycle is called ignition interrupt. In this rigorous, test the engine is cranked, and the ignition system is turned on for 1/10th of a second and off for 4/10ths of a second. This process is repeated ten times before constant ignition is applied. Once full engine speed is reached, the starter is maintained for an additional second, and the engine is then turned off for 30 seconds. The entire cycle is repeated. According to the manufacturer, only starters of the highest quality withstand this test.

The fourth cycle monitors a running engine for correct engine speed, oil pressure, and other parameters. The fifth and last check is designed for diesel engines. This unique test checks for starter and alternator durability.

The second test cell monitors the operation of engine alternators. In this test, each of the Mini-PLC-2 controllers monitors an engine and transfers pertinent data to the PLC, via the Data Highway. The PLC generates reports for the alternator and engine performance.

When an operator sends a command from the PLC controller to start testing, a Mini-PLC-2 transfers a BCD value to an analog subsystem, which tells the engine to accelerate or decelerate to a specified speed. Any variation from the preset speed causes the engine to drop to idle until the error is corrected. The Mini-PLC-2

Alternator test room controlled by independent controllers communicating through a data highway (Courtesy of Allen-Bradley Co.).

monitors key engine parameters such as alternator voltage and current. If the engine fails the test (10% deviation from specifications), it is automatically shut down. A message indicating the cause of the engine failure, along with test values, is sent over the data highway to the PLC, which generates a message printout on a teletype.

The PLC controller is also programmed to help monitor safety functions in the test lab. Rotating engines cause extremely hazardous conditions to exist in the room during testing. Among the safety systems used is a strobe light that warns workers to vacate the room fifteen seconds prior to high-speed testing. Should anyone enter the room during testing, an infrared device monitoring the door automatically sets the engines to idle speed, and a new test sequence must again be requested from the keyboard in the control room.

**Results:** With the PLC and Mini-PLC-2 controllers in conjunction with the data highway, the systems engineers were able to design a reliable engine component test system, reduce programming costs, simplify operator communications, and still permit extensive test data and error reports to be generated for management use.

**Application #15:** AUTOMATION OF ANTENNA TRANSPORT (Square D Co.)

**Controller:** SY/MAX-20

**Why PC Was Applied:** The control system needed to be flexible and easily maintained.

**Description:** The application of space technology to Earth physics may provide a means for long-range earthquake predictions; this research was put into action at a branch of the broad ranged NASA Crustal Dynamics Project. The research project involves the rotation of sensitive dish-shaped antenna receivers to receive signals from the edge of the known universe. Of the antennas used, one is mounted on a recently developed, highly mobile, 40-foot long transport trailer. The electronic system on this antenna transport is a SY/MAX-20 programmable con-

Antenna system fully deployed (Courtesy of Square D Co.).

SY/MAX 20 control system located at the front end of the trailer (Courtesy of Square D Co.).

troller, used for control of antenna deployment, storage, and trailer platform leveling. The PC is housed in the control cabinet at the front end of the trailer along with the operator station.

Upon reaching a pre-selected site under operator command, the PC controls the hydraulic and electrical systems used to open and close the roof, raise and lower the sides of the trailer, assemble and stow the antenna, and level the platform.

For antenna positioning and dish-half installation and removal, information is exchanged between the computer system and the PC through a SY/MAX computer interface module. The computer and terminal are housed in a van which accompanies the antenna transport. The electrical power for the antenna control system is completely self-contained and is provided by two generators.

**Results:** The programmable control system made it possible to assemble and disassemble the antenna system in less than 30 minutes by a two man-crew.

**Application #16:** LIQUID DYE BATCHING FOR CONTINUOUS RANGES
(Reliance Electric, Toledo Scale Systems Div.)

**Controller:** AutoMate 35

**Why PC Was Applied:** The system was needed to position accurately a dual weight hopper trolley car and to weigh various liquid dyes as required by any particular color formulation.

**Description:** The AutoMate 35 is utilized as the system controller. A variable speed motor drives the scale hopper trolley to the desired position under a dye tank or over a mix station. A position encoder with control room display transmits the position of the trolley. The small (minor) hopper allows better resolution for small weighings within each formulation. The weights are displayed in the control room by digital indicators. A manual operator's panel provides a backup mode of operation in the control room. A video terminal is utilized as the primary operator device for entering formulas and starting batches. A printer is used to generate alarm messages as well as production and material usage reports.

The control program allows the operator to modify or delete existing formulations or to define new ones. Another routine is used to start a batch or series of batches. Once a batch is initiated, the trolley is driven by the controller to position the major or minor scale hopper under the appropriate dye tank. Acceleration and deceleration is controlled to prevent sloshing of the hopper's contents and surges on the motor. From each dye tank, the addition is made at a fast rate, then switched to a slow rate for accuracy. When the batch is complete, the trolley is driven into position over the specified mix tank, and the hoppers are dumped and flushed for a timed period. The next batch is then automatically started if one has been previously specified. A pickup is located near the center of the trolley rails' span to check the position encoder each time the trolley passes. An alarm is generated if the encoder is out of tolerance.

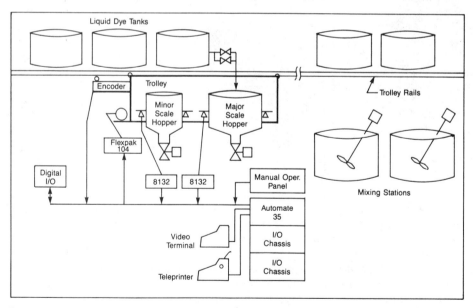

Batching process block diagram for the making of liquid dye (Courtesy of Reliance Electric, Toledo Scale Systems Div.).

**Results:** Formulations are very consistent from batch to batch due to automation and scale accuracy. Batching is much faster since there is no manual intervention. The system operator has time available to perform other dyehouse laboratory or production functions. Skill levels required of the drug room personnel are reduced. Malfunctions are displayed on digital indicators, and maintenance, batch, and summary reports provide useful data for making production and maintenance decisions.

**Application #17:** TURBINE/GENERATOR CONTROL (General Electric Co.)

**Controller:** SERIES 6 MODEL 60

**Why PC Was Applied:** The system was installed to provide a more reliable and cost effective control system for a hydrogenerator plant.

**Description:** In this installation, the Model 60 controller performs sequencing and automatic controlling of three fixed hydraulic turbines, and selects the best combination of turbine/generators to accommodate the water available.

The controller uses 2K memory, 15 AC inputs and 10 AC outputs per turbine/generator, two TTL cards, and two analog cards. The inputs and outputs are 115 VAC contact closures, (e.g., limit switches on valves) or analog devices, which give a 4-20 milliamps signal proportional to water flow.

Water level or flow is continuously monitored to see that all available generators are on line in order to produce maximum energy. If not enough generators are on line, the PC will add additional units. The PC is programmed to make use of the best combination of available turbine/generator sets for each water condition. For example, the largest turbine/generator has the best efficiency and should be on-line if water is available to run it at rated flow. If water conditions decrease, the controller will shut down the least efficient generator.

Any change in water flow, such as when another turbine comes on-line, will cause the PC to initiate an alarm to forewarn downstream fisherman of impending water flow changes. Daily production information for each generator is monitored and stored. The data gathered includes generator on/off cycles, utility outages, and generator overcurrent conditions. Any signal indicating a fault current causes an emergency turbine/generator shutdown.

Daily and monthly reports can be remotely data-logged or recorded on a printer at the powerhouse. Each unit's daily production can be monitored and compared to daily production from the previous month to determine efficiency loss trends.

**Results:** By using three turbines properly sequenced by the controller instead of a single large variable hydraulic turbine, a capital cost savings of up to 50% was achieved.

**Application #18:** CONTROL OF UNDERGROUND CHEMICALS
(General Electric Co.)

**Controller:** SERIES 6 MODEL 600

**Why PC Was Applied:** The system was needed to provide report generation and to control pump-over pressure as a safety measure.

**Description:** The Model 600 programmable controller was applied to control the flow, velocity, specific gravity, temperature, and pressure of underground chemicals. This data comes from sensors that measure the temperature density and flow rates for mass flow computations performed by the controller. Report data generation on the status of the system is provided daily or upon request.

The PC controls the relief of gases from chemicals stored under constant pressure in underground wells. If pressure builds up, it is relieved by drawing gases off to a flare stack for burning. The controller also alerts the operator if the pressure drops below 1200 psi or goes above 1325 psi, so that corrective action can take place. The controller is also used to monitor the discharge steam from pump turbines. The steam is recycled and used to help a steam-driven generator.

**Results:** The programmable controller system provided safety benefits by monitoring and controlling the underground gases and provided energy savings by recycling the steam. The cost of recycling the steam is about one-twentieth of what it would cost to buy.

**Application #19:** STONE PACKAGING SYSTEM (Courtesy of Stone Container Corp. and Omron Electronics)

**Controller:** SYS-MAC P0R

**Why PC Was Applied:** The controller was needed for OEM use for automatic control of a packaging machine.

**Description:** The OMRON POR programmable controller is used to perform the control functions of Stone's SPM101 packaging machine. In the automatic mode, the PC supervises the automatic loading of multiple products. The products are skinned and then moved onto the slitter for automatic two-way separation of the products into individual packages. The PC monitors a heat sensor to provide optimum film temperature control.

The POR system replaced relay logic and eliminated 19 relays and 4 timers for a much improved, trouble-free operation. LED indicator lights display each function so that the operator can see the cycle sequence being performed.

**Results:** The programmable controller provides smooth and effective control of the machine, reduces space, and provides automatic in-feed transport which eliminates the need for repositioning pre-loaded products before skinning. The use of this controller also led Stone Corp. to use PCs in other automatic and semi-automatic packaging machines (Case Former and Sealer, Wrap Around Caser, and others), thus positioning Stone as one of the leaders in automatic packaging machinery.

Stone's packaging machine (Courtesy of Stone Container).

Closer look at the Omron controller (Courtesy of Omron Electronics).

# CHAPTER
# --]13[--

# SELECTING THE RIGHT PROGRAMMABLE CONTROLLER

*The developments in programmable controllers have been pushed rapidly forward by new inexpensive technology, greater demands from users, and ever-increasing competition.*

## 13-1  INTRODUCTION

Programmable controllers are available in all shapes and sizes, covering a wide spectrum of capability. On the low end are *relay replacers*, with minimum I/O and memory capability. At the high end are *large supervisory controllers*, which play an important role in hierarchal systems by performing a variety of control and data acquisition functions. In between these two extremes are *multifunctional controllers* with communication capability, which allows integration with various peripherals, and expansion capability, which allows the product to grow as the application requirements change.

Deciding on the right controller for a given application has become increasingly difficult. Having been complicated by an explosion of new products, including general and special purpose programmable controllers, this process places an even greater demand on the system designer to take a system approach to selecting the best product for each task. Many factors are affected by programmable controller selection; it is up to the designer to determine what characteristics are desirable in the control system and which controller best fits the present and future needs of the application.

This chapter covers the range of PC capabilities, several guidelines for defining and configuring the control system, and also other factors that will affect the final selection.

## 13-2  A LOOK AT PC PRODUCT RANGES

Prior to evaluating the system requirements, it might be helpful to understand the various ranges of programmable controller products and typical features found within those ranges. This understanding will enable the user to identify quickly the general area in which the specifications for an application may be found and to select the product that comes closest to matching these requirements.

Figure 13-1 illustrates the product ranges, divided into four major areas showing overlapping boundaries. The basis for segmentation of the product areas is the

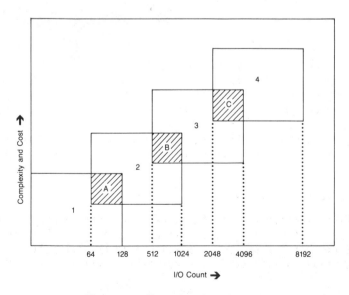

**Figure 13-1.** Illustration of product ranges.

272

number of possible inputs and outputs the system can accommodate (I/O count), the amount of memory available for the application program, and the system's general hardware and software structure. As I/O count increases, the complexity and cost of the system also increases. Similarly, as the system complexity increases, so does the memory capacity, the variety of I/O modules, and the capabilities of the instruction set.

Shaded areas A,B, and C reflect the possibility of a controller with enhanced (not standard) features for a particular range. These enhancements place the product in a grey area that overlaps the next higher range. For example, because of its I/O count, a small PC falls into area 1, but it has analog control functions that are usually standard in medium size controllers. This fact will place that product in area A. Products that fall into these overlapping areas will always allow the user to select the product that matches the requirements closest, without having to select the larger product unless it is necessary.

The typical characteristics of the various product segments are explained in the following discussions.

## Segment 1 — Small PCs

These small controllers, including Micro PCs, are usually found in applications in which logic sequencing and timing functions are required for ON/OFF control. Micro-controllers and small PCs are widely used for individual control of small machines. Often, these products are single board controllers (see Fig. 13-2). Table 13-1 lists standard features found in segment 1.

**Table 13-1.** Typical standard features for small PCs

---

- Up to 128 I/O
- 4 or 8 bit processor
- Relay replacing only
- Memory up to 2K words
- Digital I/O
- Local I/O only
- Ladder or Boolean language only
- Timers/Counters/Shift registers (TCS)
- Master Control Relays (MCR)
- Drum Timers or Sequencers
- Generally programmed with hand-held programmer

---

**Area A.** This shaded area includes controllers that are capable of having up to 64 or 128 I/O and also includes products having features normally found in medium sized controllers. The enhanced capabilities of these small controllers allow them to be used effectively in applications in which a small number of I/O is needed, yet analog control, basic math, LANs, remote I/O, and/or limited data handling may be required (Fig. 13-3). A typical case is a transfer line application in which several small machines, under individual control, must be interlocked (through LAN).

## Segment 2 — Medium PCs

Medium PCs (Fig.13-4) are applied when more than 128 I/O, analog control, data manipulation, and arithmetic capabilities are required. In general, the controllers in segment 2 are characterized by more flexible hardware and software features than those previously mentioned. These features are listed in Table 13-2.

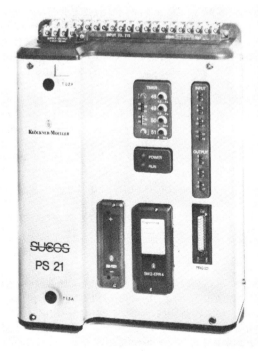

**Figure 13-2.** Klockner-Moeller Sucos P21 programmable controller, with capacity of 16 inputs and 12 outputs.

**Figure 13-3.** The Gould-Modicon Micro-84 PC with capacity of up to 64 I/O with LAN and analog capabilities.

**Area B.** In general, the shaded area B contains products in segment 2 that have more memory, table handling, PID, subroutine capability, and more arithmetic or data handling instructions. These feature are typically standard in segment 3 and basically are enhanced instruction sets. The Allen-Bradley PLC 2/30, shown in Fig. 13-5, falls into this category.

## Segment 3 — Large PCs

Large controllers (Fig. 13-6) are used in more complicated control tasks that require extensive data manipulation, data acquisition, and reporting. Further software enhancements allow these products to perform more complex numerical computations. These standard features are summarized in Table 13-3.

**Figure 13-4.** Westinghouse Numa-Logic PC-700 with capacity of 512 I/O.

**Figure 13-5.** The PLC-2/30 from Allen-Bradley with 896 I/O capacity.

**Table 13-2.** Typical standard features for medium size PCs

- Up to 1024 I/O
- 8 bit processor
- Relay replacing and analog control
- Typical memory up to 4K words. Expandable to 8K.
- Digital I/O
- Analog I/O
- Local and remote I/O
- Ladder or Boolean language
- Functional block/high level language
- TCSs
- MCRs
- Jump
- Drum Timers or Sequencers
- Math Capabilities
  —Addition
  —Subtraction
  —Multiplication
  —Division
- Limited data handling
  —Compare
  —Data conversion
  —Move register/file
  —Matrix functions
- Special function I/O modules
- RS 232 communication port
- Local Area Networks (LANs)
- CRT programmer

**Table 13-3.** Typical standard features for large PCs

- Up to 2048 I/O
- 8 or 16 bit processor
- Relay replacing and analog control
- Typical memory up to 12K words. Expandable to 32K.
- Digital I/O
- Analog I/O
- Local and remote I/O
- Ladder or Boolean language
- Functional block/high level language
- TCSs
- MCRs
- Jump
- Subroutines, interrupts
- Drum Timers or Sequencers
- Math Capabilities
  —Addition
  —Subtraction
  —Multiplication
  —Division
  —Square root
  —Double precision
- Extended Data Handling
  —Compare
  —Data conversion
  —Move register/file
  —Matrix functions
  —Block transfer
  —Binary tables
  —ASCII tables
- Special function I/O modules
- PID modules or system software PID
- One or more RS 232 communication ports
- Local Area Networks (LANs)
- Host computer communication modules
- CRT programmer

**Figure 13-6.** General Electric's Series Six model 600 with capacity of 2000 I/O.

**Area C.** The shaded area C may include segment 3 PCs that have a larger amount of application memory and more I/O capacity. Greater math and data handling capabilities can also be found in this area. The Giddings and Lewis PC-409, shown in Fig. 13-7, is an example of a controller in this area.

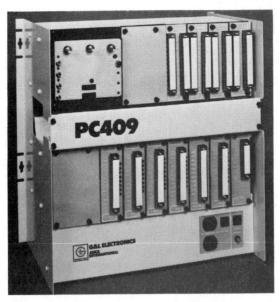

**Figure 13-7.** Gidding & Lewis PC-409 (2032 I/O adn 40K memory) can be placed in Area C for its capabilities to perform advanced axis control.

## Segment 4 — Very Large PCs

Very large PCs (Fig. 13-8) are utilized in sophisticated control and data acquisition applications when basic requirements are large memory and I/O capacity. Remote and special I/O interfaces are also a standard requirement.

Typical application areas include steel mills, paper mills, and refineries. These PCs also serve as supervisory controllers in large distributed control applications. Table 13-4 lists standard features found in segment 4.

**Figure 13-8.** The Gould-Modicon 584 PC with capability of handling up to 4096 I/O and advance programming software.

277

**Table 13-4.** Typical standard features for very large PCs

- Up to 8192 I/O
- 16 bit processor or multi-processors
- Relay replacing and analog control
- Typical memory up to 64K words. Expandable to 128K.
- Digital I/O
- Analog I/O
- Remote analog I/O
- Remote special modules
- Local and remote I/O
- Ladder or Boolean language
- Functional block/high level language
- TCSs
- MCRs
- Jump
- Subroutines, interrupts
- Drum Timers or Sequencers
- Math Capabilities
  —Addition
  —Subtraction
  —Multiplication
  —Division
  —Square root
  —Double precision
  —Floating point
  —Cosine functions
- Powerful Data Handling
  —Compare
  —Data conversion
  —Move register/file
  —Matrix functions
  —Block transfer
  —Binary tables
  —ASCII tables
  —LIFO
  —FIFO
- Special function I/O modules
- PID modules or system software PID
- Two or more RS 232 communication ports
- Local Area Networks (LANs)
- Host computer communication modules
- Machine diagnostics
- CRT programmer

## 13-3  DEFINING THE CONTROL SYSTEM

Selecting the right programmable controller to control a machine or process requires several preliminary considerations regarding not only current needs of the application, but perhaps needs that will involve future plant goals. Keeping the future in mind will allow changes and additions to the system at minimal cost. With proper considerations, the need for memory expansion may only require that a memory module be installed; the need for an additional peripheral can be accommodated if the communication port is available. A local area network consideration will allow future integration of each controller into a plantwide communication scheme. If present and future goals are not properly evaluated, the control system may quickly become inadequate and obsolete.

Once basic considerations have been investigated, you should be prepared to begin defining specific controller requirements. The following items need to be evaluated and defined:

- Input/Output
- Type of Control
- Memory
- Software
- Peripherals
- Physical and Environmental

## I/O Considerations

Determining the amount of I/O is typically the first problem that must be addressed. Once the decision has been made to automate some machine or process either totally or partially, determining the amount of I/O is simply a matter of counting the discrete and/or analog devices that will be monitored or controlled. This determination will help to identify the minimum size constraints for the controller. Remember to allow for future expansion and spares (typically 10% to 20% spares). Spares do not effect the choice of PC size.

**Discrete Inputs/Outputs.** Input/Output interfaces of standard ratings are available for accepting signals from sensors and switches (e.g.pushbuttons, limit switches, etc.) as well as control (ON/OFF) devices (e.g. pilot lights, alarms, motor starters, etc.). Typical AC inputs/outputs range from 24 volts to 240 volts, and DC inputs/outputs from 5 volts to 240 volts.

Although input circuits vary from one manufacturer to another, certain characteristics are desirable. Look for debouncing circuitry to protect against false signals and surge protection to guard against large transients. Most input circuits will have optical or transformer isolation between the high power input and the control logic circuitry of the interface circuit.

When evaluating discrete outputs, look for the following: fuses, transient surge protection, and isolation between the power and logic circuits. Fused circuits usually cost more initially, but will probably cost less than having the fuse installed externally. Check for fuse accessibility; replacing fuses may mean shutting down several other devices. Most output circuits with fuses will also have blown fuse indicators, but make sure to check. Finally, check the output current ratings and the specified operating temperature. Typically, the temperature rating is at 60°C.

Interface circuits with isolated commons (return lines) will be required if the input/output devices are powered from separate sources.

**Analog I/O.** Analog input/output interfaces are used to sense signals generated from transducers. These interfaces measure quantity values such as flow, temperature, and pressure and are used for controlling voltage or current output devices. Typical interface ratings include -10V to +10V, 0V to +10V, 4 to 20mA, or 10 to 50mA.

Some manufacturers provide special analog interfaces to accept low level signals (e.g. RTD, thermocouple). Typically, these interface modules will accept a mix of various thermocouple or RTD types on a single module. Users should consult the vendor concerning specific requirements.

**Special I/O.** When selecting a programmable controller, the user may be faced with a situation requiring some special type of I/O conditioning (e.g. positioning, fast input, frequency, etc.) that may be impossible to implement using standard I/O. The user should find out whether or not the vendors under consideration have special modules that would help minimize control efforts. Smart modules, a subset of special interfaces, should also be considered. Typically, these interfaces perform all the field data processing in the module itself, thus relieving the CPU from performing time consuming tasks. For example, PID, 3 axis positioning, and stepper motor control modules would make the control implementation much easier and feasible, thus reducing programming and implementation time.

**Remote I/O.** Remote I/O should always be considered, especially in large systems. I/O subsystems, located a distance (over twisted pairs) away from the CPU, can mean a dramatic reduction in wiring cost, both from a labor and material standpoint. Another advantage of remote I/O subsystems is that the inputs/outputs can be strategically grouped to control separate machines or sections of a machine or process. This grouping allows easy maintenance and start-up without involving the entire system. Most controllers that have remote I/O also have remote digital I/O. However, users who require remote analog I/O should check to see if it is available in the products being considered.

## Types of Control

With the advent of new, smarter programmable controllers, the decision on the type of control has become a very important initial step. Questions such as, *what type of control should I use?* are more often asked when automating a process. Knowing process application and future automation requirements will help the user to decide what type of control will be required. This knowledge will ease the selection of a particular PC. Possible control configurations include *individual* (or *segregated*) *control, centralized control,* or *distributed control.* Figure 13-9 illustrates these configurations.

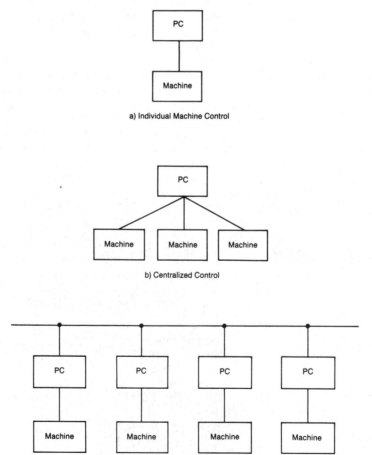

a) Individual Machine Control

b) Centralized Control

c) Distributed Control

**Figure 13-9.** Control configuration.

Individual or segregated control is used when a PC controls a single machine with only local I/O or with local and a few remote I/O. This type of control does not normally require communication with any other controllers or computers. Individual control is primarily applied to OEM and end-user equipment, such as injection molding machines, small machine tools, and small dedicated batching processes. When deciding on this approach, the user should take into consideration whether future intercontroller communication will be desired. If so, the appropriate controller can be selected then to avoid extra design expenses at a later date.

Centralized control is used when several machines or processes are controlled with one central programmable controller. This type of control can have many subsystems spread out through the factory, each one interfaced with specific I/O devices that may or may not be related to the same control. The communication that exists is only to subsystems and/or peripherals; no exchange of PC status or data is sent to other PCs.

The flexibility and potential of these applications depend greatly on the PC used and the design philosophy of the system's designer. Centralized control can be interpreted as a large individual control applied to a large process or machine. Some processes require central control due to the complexity of decentralizing the control tasks into smaller ones. One distinctive disadvantage of centralized control is that if the main PC fails, the whole process is stopped. In critical, large central control, when a back-up is needed, redundant systems can be used to overcome this problem. Several manufacturers offer this redundancy option.

Distributed control involves two or more PCs communicating with each other to accomplish the complete control task. This type of control typically employs Local Area Network (LAN), in which several PCs are controlling different stages or processes locally and are constantly interchanging information and status regarding the process. Communication among PCs is done through single coaxial cables or fiber-optics at very high speeds (up to 1 Megabaud). Despite this powerful configuration, communication between two different manufacturer's LAN systems can be difficult. Therefore, the user should define properly from the beginning the functional needs of the PCs to be used according to the application.

## Memory Considerations

The two main considerations here are the type and the amount of memory. The application may require both nonvolatile memory (i.e. retains contents upon loss of power) and volatile memory with battery backup. A nonvolatile memory such as EPROM will provide a reliable permanent storage medium, once the program has been created and debugged. If the application will require on-line changes, Read/Write memory supported by battery should be considered. Some controllers will offer both options to be used singularly or in conjunction with each other.

The memory capacity of small PCs is normally fixed (non-expandable) between 1/2 K and 2K words and is, therefore, not really a major selection concern. In medium and large controllers, however, memory can be expanded incrementally in units of 1K, 2K, 4K, etc. Although there are no fixed rules for determining the amount of memory, certain guidelines can be used to evaluate memory requirement.

The amount of memory required for a given application is a function of the total number of inputs and outputs to be controlled and the complexity of the control program. The complexity refers to the amount and type of arithmetic and data manipulation functions that will be performed. Manufacturers will normally have a rule-of-thumb formula for each of their products that will aid in making an approximation of the memory requirement. This formula involves multiplying the total number of I/O by some constant (usually a number between 3 and 8). An additional 25% to 50% should be added to the first approximation, if the program involves arithmetic or data manipulation.

Finally, the best way to obtain the memory requirement is to create the program and count the number of words used. Knowledge of the number of words required to store each instruction will allow the user to determine exact memory requirements.

## Software Considerations

During implementation of the system, the user will be faced with the programming of the PC. Because the programming is so important, users should be aware of the software capabilities of the product they choose. Generally, the software capability of a system is tailored to the handling of the control hardware that is available with the controller. However, an application may require special software functions beyond the control of the hardware components. For instance, an application may involve special control or data acquisition functions that require complex numerical calculations and data handling manipulations. The instruction set selected will determine the ease with which the software task can be implemented. The available instruction set will directly affect the time needed to implement the control program and the program execution time.

## Peripherals

The first peripheral to consider is a programming device. In general, programmers are available in two types the small (low cost) handheld units or CRT display type. The handheld programmer, typically used with small PCs, can display either a single program element or in some cases a single program rung.

Both handheld and CRT programmers provide programming and monitoring capability, but the CRT allows greater flexibility. Some CRT programmers also provide program storage capability, as well as limited graphics. Intelligent CRTs, as they are labeled, are microprocessor based devices that can be programmed for various functions. The capability of these units differs depending on the manufacturer. Typical features are listed later in this section.

In addition to the programming device, users may require peripherals at local or remote control stations to provide interface between the controller and an operator. Two of the most common peripherals are the line printer, used for obtaining a hardcopy printout of the program, and a cassette recorder, used to store the program. Messages or alarms can be sent to graphic or alphanumeric display devices, while hourly or monthly production reports can de stored on a floppy diskette and later sent to a line printer.

Peripheral requirements should be evaluated along with the CPU, since the CPU will determine the type and number of peripherals that can be interfaced, as well as the method of interfacing. The distance that peripherals can be placed from the CPU is also affected. Typical peripherals are printers, CRTs, diskette drives, color displays, alphanumeric displays, and cassette drives.

## Physical and Environmental

The physical and environmental characteristics of the various controller components will have significant impact on the total system reliability and maintainability. Consider the ambient conditions in which the controller will be operating. Conditions such as temperature, humidity, dust level, and corrosion can all affect the controller's ability to operate properly. It is important to determine operating parameters (i.e. temperature, vibration, EMI/RFI, etc.), and packaging (i.e. dust proof, drip proof, ruggedness, type of connections, etc.) when specifying the controller and its I/O system. Most programmable controller manufacturers provide products that have undergone certain environmental and physical tests (e.g.

temperature, EMI/RFI, shock, vibration, etc.). Users should be aware of the tests performed and whether or not the results meet the demands of the operating environment.

Checklist with typical sample answers are given in Table 13-5. This list will help to examine most of the features a user should look for when evaluating PC requirements. Note that all product ranges are covered in the list, from small to very large. Some PCs may not answer all checks, primarily due to their range characteristics.

**Table 13-5.** System Checklists

| I/O SYSTEM CHECKLIST | TYPICAL ANSWERS |
|---|---|
| **I/O Count** | |
| • Digital count | Maximum of 128 I/O mixable |
| • Analog count | Maximum of 16 I/O mixable |
| **Digital I/O** | |
| • Inputs | |
|     Number of points/module | 4 points/module, etc. |
|     Input Type | AC,DC,non-voltage, etc. |
|     Input ratings | 110 VAC, 220 VAC, 5-24 VDC, contact, etc. |
|     Maximum inputs/channel | 64 points/channel |
|     Input status indicators | Power, logic |
|     Isolation | 1500 Volts optical |
| • Outputs | |
|     Number of points/module | 16 points/module |
|     Output type | AC,DC, contact, etc. |
|     Output ratings | 110 VAC, 220 VAC, contact |
|     Output current Amps/point | 1 Amp/point with all outputs on @ 115 VAC |
|     Maximum outputs/channel | 64 points/channel |
|     Output status indicators | Power, individual blown fuse, logic |
|     Output protection | Fuse, suppression on contact output |
| **Analog I/O** | |
| • Inputs | |
|     Number of points/module | 4 analog inputs/module |
|     Resolution | 11 bits |
|     Input type | current, voltage |
|     Input ratings | 4-20 ma., 0-5 volts, 0-10 volts |
|     Built-in transducer | Yes, thermocoupler input |
|     Maximum analog inputs/channel | 32 |
|     Supply of power | Internal to PC |
| • Outputs | |
|     Number of points/modules | 2 analog outputs/module |
|     Output type | current, voltage |
|     Output ratings | 4-20 ma., 10-50 ma., 0-10 volts |
|     Maximum analog outputs/channel | 16 |
|     Supply of power | +15 VDC and -15 VDC |
| **Remote I/O** | |
| • Digital | |
|     Distance | 1500 ft. |
|     Number of I/O per remote | 32 I/O per remote |
|     Communication Link | Twisted pair, 100 ohms impedance |
| • Analog | |
|     Distance | 5000 ft. with receiver/transmitter |
|     Number of I/O per remote | 16 analog inputs/outputs per remote |
|     Communication Link | coaxial |
| **Special I/O** | |
| • High-speed pulse counter | Local and remote, 50 KHz |
| • Fast electronic input | Yes, 5 microsecond pulse width minimum |
| • Interrupt module | Yes |
| • Absolute encoder | Direct connection to encoder |
| • Incremental encoder | Not available |
| • BCD input/output module | Remote, 4 and 8 BCD digits |

**Table 13-5.** System Checklist **(Continued)**

| | |
|---|---|
| • Stepper motor | Yes |
| • ASCII communications module | Yes, full ASCII, 300-4800 baud |
| • Host computer | Yes, protocol decode on module |
| • LAN I/O module | Extra board in CPU |
| • PID module | Local and remote, 2 loops per module |
| • Language module | Yes, basic interpreter module |
| **PHYSICAL** | |
| • Wire size to I/O | 20 AWG, can handle two wires per I/O |
| • Separate commons | Yes for 4 points/module, no for 16 |
| • Removable under power | Yes |
| • Disturb wiring to remove I/O | Disconnect screw from I/O module |

| CENTRAL PROCESSING UNIT CHECKLIST | TYPICAL ANSWERS |
|---|---|
| **Processor** | |
| • Microprocessor | 8 bit micro, and multiprocessor board |
| • Scan time | 10 msec/K of memory |
| • Communication ports | Two RS-232C ports |
| **Memory** | |
| • Memory type | RAM, EEPROM |
| • Total system memory | 64K |
| • Application memory size | 8K for user |
| • Word size | 8 bits |
| • Memory utilization | 1 word per element (coil, contact) |
| • Memory protect | Yes, key switch |
| **Power Supply** | |
| • Incoming Power | 120/240 VAC, 24 VDC |
| • Frequency | 50/60 Hz |
| • Voltage variation | +15%, -10% |
| • Overvoltage protection | Yes |
| • Current limiting | Yes |
| • Maximum current supply | 100 ma. @ 14 VDC, 2.5 amps @ 5 VDC |
| • Isolation | 1500 volts |
| • Location    Built-in CPU | |
| **Environmental** | |
| • Operating temperature | 0-60 degrees C |
| • Humidity | 5%-90% relative humidity non-condensing |
| • EMI/RFI | Satisfies NEMA and IEEE tests |
| • Noise | ,000 volts peak-peak, 1 microsecond |
| • Vibration | Withstands 16.7 Hz, double amplitude |
| • Shock | 10g X,Y, and Z direction |

| SOFTWARE CHECKLIST | TYPICAL ANSWERS |
|---|---|
| **Language** | |
| • Ladder or Boolean | Ladder language |
| • High level | Functional blocks |
| **Software Coils** | |
| • Number of internals | 128 |
| • Number of timers | 32 timers, maximum count 9999 sec. BCD |
| • Number of counters | 166 |
| • Number of shift registers | 32, 16 bit each |
| • Number of drum timers | 16 |
| • Timer's time base | 0.1, 1.0 seconds |
| • Timer type | on delay and off delay |
| • Counter type | up/down count |
| • Latch coil | 32 |

**Table 13-5.** System Checklist **(Continued)**

| | |
|---|---|
| • Transitional coil | 16, OFF-ON, and ON-OFF |
| • MCR | 8 |
| • Global coil | 256 in LAN |
| • Global register | 128 in LAN |
| • Fault coil | Yes, detection of CPU failure |
| • Interrupt coil | Yes |
| **Math** | |
| • Addition | Yes, double precision |
| • Subtraction | Yes, double precision |
| • Multiplication | Yes |
| • Division | Yes |
| • Square root | Yes |
| • Floating point | Yes, $1E+38$, $1E-38$ |
| • Trigonometric functions | Yes, sine and cosine |
| **Data Handling** | |
| • Number of registers | 128, 16 bits each |
| • Data size in registers | $+32767$, $-32767$, and 9999 BCD |
| • Compare | Yes, greater/less than or equal to |
| • Conversions | Binary-BCD, BCD-Binary |
| • Move | Registers and single files |
| • Matrix | AND, OR, EXOR, NAND |
| • Tables | Move to ASCII or binary tables |
| • PID | Software functional block, 20 loops |
| • LIFO | Yes |
| • FIFO | Yes |
| • Jump | Conditional and direct |
| • Subroutines | Yes |
| • ASCII instructions | Yes, print and read |
| • Sort | No |
| • Machine diagnostics | Yes |

| PROGRAMMING AND STORAGE DEVICES CHECKLIST | TYPICAL ANSWERS |
|---|---|
| **CRT Or Computer** | |
| • Physical | |
|     CRT or computer | CRT |
|     Display size | 9" screen |
|     Graphics | No |
|     Ladder matrix size | 10x7 elements |
|     Built-in storage | Yes, tape |
|     LAN | No |
|     Communication | RS-232C and 20 ma current loop |
|     Incoming power | 115/230 VAC |
|     Operating temperature | 0-40 degrees C |
|     Keyboard type | Mylar or standard keys |
| • Functional | |
|     Intelligent Yes | |
|     Single scan | No |
|     Power flow | Yes, element intensified on screen |
|     OFF-LINE programming | Yes |
|     Monitor function | Yes |
|     Modify function | Yes |
|     Force I/O | Yes, indicates forcing on mainframe |
|     Search | No |
|     Mnemonics | Yes |

**Table 13-5.** System Checklist **(Continued)**

**Manual Programmer**
- Physical
  - Display type — LCD or LED
  - Ladder matrix size — 7x4 elements
  - Communication — RS-232C
  - Incoming power — From unit
  - Operating temperature — 0-40 degrees C
  - Keyboard type — Mylar
- Functional
  - Intelligent — No
  - Single scan — No
  - Power flow — Yes
  - Monitor function — Yes
  - Modify function — Yes
  - Force I/O — Yes
  - Search — No
  - Mnemonics — Yes, also messages

**Storage Devices**
- Digital cassette — Yes, STR-LINK
- Floppy disk — No
- Computer — Yes, thru computer module
- Electronic memory module — Yes, for small PC

| SYSTEM DIAGNOSTICS CHECKLIST | TYPICAL ANSWERS |
|---|---|
| **Power Supply** | |
| • Power loss detection | Yes, after 3 cycles |
| • Voltage level detection | Yes, DC levels for CPU |
| • Diagnostic monitoring | Continuously |
| **Memory** | |
| • Memory OK | Yes, checksum, LED indicator |
| • Battery OK | Yes, LED indicator |
| • Diagnostic monitoring | At power up only |
| **Processor** | |
| • Local | Yes, watch dog timer, and LED indicator |
| • Remote | Yes, indicator in CPU |
| • Diagnostic monitoring | Continuously |
| **Communication** | |
| • Local I/O | Yes |
| • Remote I/O | Yes, check-sum |
| • Programming device | CRT port OK, and RS-232C OK |
| • Diagnostic monitoring | During transmission |
| **Fault Indications** | |
| • CPU | Yes, external relay contacts |
| • Remote | Yes, external contacts at remote driver |
| • LAN | Yes, internal coil |
| • I/O | Yes, detects presence of I/O module |

## 13-4   OTHER FACTORS AFFECTING SELECTION

An evaluation of the hardware and software requirements will narrow the selection down to one of a few possible candidates. Eventually, two or more products will meet all the requirements of the preliminary system design, and then a final decision must be made. At this point, there are still several other factors that can lead to a final selection of the product that best fits the system specifications and the in-house requirements. The following considerations should be made and discussed with the potential vendors.

### Vendor Support

**Sales and Marketing.** Choosing a vendor who is willing and capable of providing good technical support can make life with a programmable controller much easier. Quality technical support can begin even before a purchase is made. A good indicator of future support is a sales and marketing organization that is close by to help with any preliminary questions the user might have. They should be able to answer any questions that sales and promotional literature do not detail.

**Training.** Another form of vendor support is training for in-house personnel. Understanding the basic elements of the product selected will allow the user to apply this knowledge effectively, and thus, obtain the best system results. Most vendors offer programming and maintenance schools at their own plant location; however, arrangements can sometimes be made to have the vendor provide training at the customer site. This latter arrangement, if possible, will allow more plant personnel to attend. Although most vendors charge for this service, it is sometimes offered at no cost. Establish these and other details prior to making a purchase decision.

**Field Service.** The third and probably the most important vendor support function is that of field service. Once a controller has been purchased, it is most likely that some type of field assistance will eventually be required. The programmer may need help during the programming phase, or the designer may need clarification of specifications on some interface. Field service may mean a phone consultation with an applications engineer or on-site assistance from a field service engineer.

Some PC vendors have sales, service, and distributor networks that can adequately serve technical needs, while some may lack this expertise. Whatever the service requirement, good prompt field service can mean the difference between dollars saved and dollars lost. An investigation of the quality of field service that can be expected from potential vendors would be helpful. Often, other users are willing to share this type of information.

**Technical Literature.** Supporting technical literature — programming, operational, and maintenance manuals — can prove to be an enormous assistance or a dreadful annoyance. An orderly and well-written user's manual minimizes not only delay, but also the possibility of costly misapplication. Most vendors will provide a user's manual on loan for review.

Other technical publications provided by some vendors include *application notes*. These technical articles often provide useful examples of hardware and software techniques, used in actual or potential applications. Some manufacturers provide application notes on a regular basis, while others provide them on request. Technical publications are another area for investigation.

**Product Availability.** The nature of PCs allows the control system to be installed almost completely before the product itself is needed. All other mechanical equipment can be installed, and all field I/O devices can be terminated prior to installing the controller equipment. This design approach is quite acceptable, if the exact availability of the product is known. Needless to say, however, almost nothing is exact. The ability to start up on time depends on whether or not the vendor can deliver the product within the time constraints.

The availability of parts in emergency situations is another important consideration. Such a situation exists if a failure occurs and a replacement part is not in plant inventory. The downtime in this situation can be minimized if the vendor has fast delivery from the factory or a local stocking distributor.

## Product Proven Reliability

The reliability of the controller plays an important role in overall system performance. Lack of reliability usually translates into downtime, poor quality product, and higher scrap levels.

Several factors can be investigated to determine the proven reliability of a particular product. Mean Time Between Failure studies (MTBF) can be helpful if the user can evaluate the data. Knowledge of a similar application in which the product has been sucessfully applied is also useful. A sales representative can provide such information or even on occasion arrange an on-site visit. If there are any unique or peculiar specifications that must be met (e.g. EMI and vibration specs), be sure that the vendor can truly satisfy these requirements. Finally, burn-in procedures of the product should be investigated (e.g. total system burn-in process or parts burn-in process). In general, any of this information can be obtained from the vendor upon request.

## Plant Standardization

A final consideration that may help to finalize the decision on a product is the possibility of future plant goals to standardize on a given manufacturer and product line. This route is being taken by several companies today for many good reasons. The fact that several vendors are creating complete product families of PCs that cover the entire range of capabilities is making standardization more feasible.

A current trend by manufacturers is to build completely intercompatible product families, with products ranging from the very small to the very large. These families share the same I/O structure, programming device, and elementary instruction set. They also have similar memory organization and structure. Most families of products can also be linked in a network configuration. Family traits, such as the following, provide several important benefits that should be considered:

- Training on a new family member is usually a progression instead of totally new training on a new product.
- Standard products can mean better plant maintenance in emergency situations.
- I/O spares can be used for all family products (mimimum spare inventory).
- An outgrown product can be replaced by the next larger product by simply removing the smaller CPU, installing the larger CPU, and reloading the old program.

## 13-5   SUMMARY

This chapter has presented a general approach for making a programmable controller selection. We have seen that this selection relates not only to such obvious factors as I/O capacity, memory capacity, or sophistication of control, but also to intangible factors that have a significant impact on final system results. The following steps are a summary of the major considerations involving PC selection.

**Step 1:** Know the process to be controlled.

**Step 2:** Determine the type of control.
a. Distributed control
b. Centralized control
c. Individual machine control

**Step 3:** Determine the I/O interface requirement.
a. Estimate digital and analog inputs and outputs
b. Check for input/output specifications
c. Determine if remote I/O is required
d. Special I/O requirements
e. Allow for future expansion

**Step 4:** Determine software language and functions.
a. Ladder, Boolean, and/or high level
b. Basic instructions (timers, counters, etc.)
c. Enhanced instructions/functions (math, PID, etc.)

**Step 5:** Consider the type of memory.
a. Volatile (R/W)
b. Non-volatile (EEPROM, EPROM, NOVRAM, Core, etc.)
c. Combination of volatile and nonvolatile

**Step 6:** Consider memory capacity.
a. Estimate memory usage based on memory utilization per instruction
b. Allow extra memory for complex programming and future expansion

**Step 7:** Evaluate processor scan time requirements.

**Step 8:** Define programming and storage device requirements.
a. CRT
b. Computer
c. Cassette or disk storage
d. Mini-programmer
e. Consider the functional capabilities of programming devices

**Step 9:** Define peripheral requirements.
a. Graphic display
b. Operator interface
c. Line printers
d. Documentation systems
e. Report generation systems

**Step 10:** Determine any physical and environmental constraints.
a. Available space for system
b. Ambient conditions

**Step 11:** Evaluate other factors that can affect selection.
a. Vendor support
b. Product proven reliability
c. Plant goals for future (e.g. standardization)

# APPENDIX

**-A-**

**ac input interface:** An input circuit that conditions various ac signals from connected devices, to logic levels that are required by the Processor. This interface may or may not be a module type. Typical voltages for ac inputs include 24vac, 115vac, and 230vac.

**ac output interface:** An output circuit that switches the User supplied control voltage required, to control connected ac devices. This interface may or may not be a module type. Typical voltages for ac outputs are 24 vac, 115 vac, and 230 vac.

**acknowledge:** A control signal that indicates acceptance of data, in an I/O transmission process. An acknowledge can be implemented with hardware or software.

**ADCCP:** Advanced Data Communication Control Procedures. An ANSI standard communications protocol for synchronous links, in which various combinations of primary and secondary link control functions are defined.

**address :** A reference number assigned to a unique memory location. Each memory location has an address and each address has a memory location.

**address field:** The sequence of eight (or any multiple of eight, if extended) bits immediately following the opening flag sequence of a frame. The address field identifies which secondary station is sending (or is designated to receive) the frame.

**addressability:** The total number of devices that can be connected to the network. The ability of a network to accomodate device expansion.

**algorithm:** A set of procedures for solving a problem. More generally used in reference to a software program.

**alphanumeric:** Character strings consisting of any combination of alphabets, numerals, and/or special characters ( e.g. A15$ ), for representing text, commands, numbers, and/or code groups.

**analog device:** Apparatus that measures continuous information (e.g. voltage-current). The measured analog signal has an infinite number of possible values. The only limitation on resolution is the accuracy of the measuring device.

**analog input interface:** An input circuit that employs an analog-to-digital converter to convert an analog value, measured by an analog measuring device, to a digital value that can be used by the Processor.

**analog output interface:** An output circuit that employs a digital-to-analog converter to convert a digital value, sent from the Processor, to an analog value that will control a connected analog device.

**analog signal:** One having the characteristic of being continuous and changing smoothly over a given range, rather than switching suddenly between certain levels as with discrete signals.

**AND:** A Boolean operation that yields a logic "1" output if all inputs are "1," and a logic "0" if any input is "0."

**ANSI:** American National Standards Institute.

**application layer:** Layer 7 and the highest layer of the Open Systems Architecture. This layer applies end-user information to the network. It serves the user by supplying information for real applications.

**application memory:** That part of the total system memory that is available for the storage of the application program and associated data.

**application program:** The set of instructions written by the User, for control, data acquisition, or, report generation. This software is stored in the application memory.

**arithmetic capability:** The ability to perform such math functions as addition, subtraction, multiplication, division, square roots, etc. A given controller may have some or all of these functions.

**arithmetic logic unit:** A processor subsystem that performs arithmetic operations like addition and subtraction, and logic operatons such as Exclusive-OR, AND, and OR functions. Also called ALU.

**assembly language:** A symbolic programming language that can be directly translated into machine language instructions.

**asynchronous response mode (arm):** A mode in which a secondary station may initiate transmission without explicit permission from the primary.

**asynchronous transmission:** Transmission controlled by start and stop elements at the beginning and end of each character. Time intervals between characters may be of unequal length.

-B-

**baseband:** Transmission of one signal at a time either modulated or unmodulated.

**baud:** Officially defined as the reciprocal of the shortest pulse width in a data communication stream, but usually used to refer to the number of binary bits transmitted per second during a serial data transmission.

**BCD:** See Binary Coded Decimal.

**Binary Coded Decimal:** A binary number system in which each decimal digit from 0 to 9 is represented by four binary digits (bits). The four positions have a weighted value of 1,2,4, and 8 respectively starting with the least significant bit. A thumbwheel switch is a BCD device, and when connected to a programmable controller, each decade requires four wires. Decimal 9 = 1001 BCD.

**Binary Number System:** A number system that uses two numerals (binary digits), "0" and "1." Each digit position for a binary number has a place value of 1,2,4,8,16,32,64,128, and so on beginning with the least significant (right-most) digit. Base 2. Example: $1101 = 1(1) + 0(2) + 1(4) + 1(8) = 13$.

**binary word:** A related grouping of ones and zeros having coded meaning assigned by position, or as a group, has some numerical value. 10010010 is an eight bit binary word , in which each bit could have coded significance or as a group represent the number 146 in decimal.

**bit:** One binary digit. The smallest unit of binary information (Abbreviation of Binary digIT). A bit can have a value of "1" or "0."

**bit-oriented protocol:** A type of data link control that uses a minimum of control characters, relying instead on bit position to determine the message. A resulting advantage is code-transparency.

294

**Boolean operators:** Logical operators such as AND, OR, NAND, NOR, NOT, and Exclusive-OR, that can be used singly or in combination to form logical statements or circuits that can have an output response be true or false.

**broadband:** Transmission of two or more channels simultaneously via frequency division multiplexing.

**buffer:** 1) In software terms a register or group of registers used for temporary storage of data, to compensate for transmission rate differences between the transmitting and receiving device. 2) In hardware terms, a circuit that restores a signal to a proper drive level.

**bugs:** 1) Software errors that cause unwanted behavior. 2) Functional problems due to hardware.

**burn-in:** The process of operating a device at an elevated temperature to improve the probability that any component weakness will cause a failure. The intent is to isolate any early-failing parts.

**bus:** 1) A group of lines used for data transmission or control. 2) Power distribution conductors.

**byte:** A group of adjacent bits usually operated upon as a unit, such as when moving data to and from memory. There are eight bits/byte.

## -C-

**cathode ray tube:** See CRT

**CCITT:** International Telegraph and Telephone Consultative Committee. International standards committee responsible for Recommendation X.25.

**Central Processing Unit:** That part of the programmable controller that governs system activites, including the interpretation and execution of programmed instructions. In general the CPU consists of the arithmetic-logic unit, timing/control circuitry, accumulator, scratch pad memory, program counter and address stack, and an instruction register. The Central Processing Unit is sometimes referred to as the Processor or the CPU.

**channel capacity:** The amount of information that can be transmitted per second on a given communications channel. A function of medium, line length and modulation rate.

**character:** One symbol of a set of elementary symbols, such as a letter of the alphabet or a decimal number. Characters may be expressed in many binary codes. An ASCII character can be represented by a group of seven or eight (with parity) bits.

**character stuffing:** In character-oriented protocols, a technique for distinguishing control characters from their corresponding bit sequences by the insertion and deletion of special control characters into the data stream.

**character-oriented protocol:** A type of data link control in which trasmitted data is encoded using an alphabet of (multi-bit) characters, such as ASCII. Compare to "bit-oriented."

**checksum:** A character placed at the end of a data block, that corresponds to the binary sum of all the characters in the block. This is one technique used for error detection.

**chip:** A very small piece of semiconductor material, on which electronic components are formed. Chips are normally made of slilicon and are typically less than 1/4 inch square and 1/100 inch thick.

**clear:** To remove data from a single memory location or all memory locations, and return to a non-programmmed state or some initial condition (normally "0").

**clock signal:** A clock pulse that is periodically generated and used throughout the system to synchronize equipment operation.

**CMOS:** Abbreviation for Complementary Metal Oxide Semiconductor. An integrated circuit family characterized by low power consumption and high noise immunity.

**coaxial cable:** A transmission line consisting of a central conductor surrounded by dielectric materials and an external conductor, and possessing a predictable characteristic impedence.

**code:** 1) A binary representation of numbers, letters, or symbols which has some meaning. 2) A set of programmed instructions.

**command:** In data communications, an instruction represented in the control field or a frame and transmitted by the primary device. It causes the addressed secondary to execute a specific data link control function.

**common carrier:** A public utility company (e.g. Bell, Western Union), that supplies communication services to the general public.

**complement:** A logical operation that inverts a signal or bit. The complement of 1 is 0, and the complement of 0 is 1.

**computer interface:** Circuitry designed to allow communication between a computer and some other processor, such as the programmable controller's Central Processing Unit.

**contact:** 1) One of the conducting parts of a relay, switch, or connector that are engaged or disengaged to open or close the associated electrical circuits. 2) In reference to software: the juncture point that provides a complete path when closed. See ladder contact.

**contact symbology:** A set of symbols used to express the control program using conventional relay symbols (e.g. -] [- normally open contact, -]/[- normally closed contact, and -( )- coil).

**contention system:** A communications network in which two or more stations have equivalent status and contend for access to the bus.

**control field:** The sequence of eight (or sixteen, if extended) bits immediately following the address field of a frame. The content of the control field is interpreted by: a) the receiving secondary, designated by the address field, as a command instructing the performance of some specific function. b) the receiving primary, as a response from the secondary, designated by the address field, to one or more commands.

**control logic:** The control plan for a given system. The program.

**core memory:** A type of memory that uses tiny magnetic rings to store each bit as a permanent magnetic field. This nonvolatile memory was utilized in many of the earlier programmable controllers.

**counter:** 1) An electro-mechanical relay-type device that counts the occurrence of some event. The event to be counted may be pulses developed from operations such as switch closures, interruptions of light beams, or other discrete events. 2) A programmable controller eliminates the need for hardware counters, by using software counters. The software counter can be given a preset count value , and will count up or down, depending on program, whenever the counted event occurs. The software counter has greater flexibility than the hardware counter.

**CPU:** See Central Processing Unit.

**CRT:** Abbreviation for cathode-ray tube, a vacuum tube with a viewing screen as an integral part of its envelope.

**CRT programmer:** A programming device containing a cathode ray tube. The CRT Programmer is primarily used to create and monitor the control program, but can also be used to display data and on-line reports. Some CRT programmers incorporate storage devices to record the control program.

**current loop:** A two wire communication link in which the presence of a 20 milliamp current level indicates a binary "1" (mark), and its absence indicates no data, a binary "0" (space).

**cursor:** An illuminated position indicator on the display of the programming device, that indicates where the next typed character will appear.

**cyclic redundancy checking (CRC):** An error detection scheme in which all the bits in a block of data are divided by a predetermined binary number. A check character is determined by the remainder.

-D-

**data highway:** The means of transmitting frames between stations interconnected by a data transmission line. A data highway consists of a data circuit and the Physical and Data Link Layers of the stations connected to the data circuit.

**data integrity:** The ability of a communication system to deliver data from its origin to its destination with an acceptable residual error rate.

**data link layer:** Layer 2 of the OSI architecture. It provides functional and procedural means to establish, maintain and release data-link connections among network-entities. A data-link-connection is built upon one or several physical-connections. The objective of this layer is to detect and possibly correct errors which may occur in the physical layer. In addition, the Data Link Layer conveys to the Network Layer the capability to request assembly of data circuits within the physical layer (i.e., the capability of performing control of circuit switching).

**Data network:** The means of transmitting messages between a data source and one or more data sinks. A data network may contain one or more data highways interconnecting the same or different sets of devices. A data network consists of these highways and the Network Layers of the stations interconnected by these highways.

**data signalling rate:** The rate, expressed in bits per second, at which data are transmitted or received by a data terminal equipment.

**data transfer:** The process of moving information from one location to another, register-to-register, device-todevice, etc.

**data transmission line:** A medium for transferring signals over a distance.

**debouncing:** The act of removing intermediate noise states from a mechanical switch.

**debug:** The process of locating and removing mistakes from a software program or from hardware interconnections.

**DCE:** Data Communications Equipment. Equipment designed to establish, maintain, and terminate a connection.

**decentralized connection:** A multi-endpoint connection in which data sent by any entity associated with a connection-endpoint is received by all other entities.

**Decimal Number System:** A number system that uses ten numeral digits (decimal digits), 0,1,2,3,4,5,6,7,8,9. Each digit position has a place value of 1,10,100,1000, and so on, beginning with the least significant (right-most) digit. Base 10.

**democratic system:** A distributed system which attempts to maintain equal access times for all stations.

**device driver:** A program that services a peripheral device by controlling its hardware activities. The interface between a device and the I/O code of a network

**diagnostic program:** A system program that checks a specific system operation, or a user application program that detects certain malfunctions of the controlled machine or process. Typical system diagnostics are CPU, Memory, Input/Output communication to processor program scan loss etc.. An application diagnostic program may detect and in some cases prevent certain machine or process failures.

**digital signal:** One having the characteristic of being discrete or discontinuous in nature. One that is present or not present, can be counted and represented directly as a numerical value.

**digital device:** One that processes discrete electrical signals.

**direct memory access (DMA):** A process in which a direct transfer of data to or from the memory of a processor-based system can take place without involving the central processing unit.

**disk cartridge:** Some diskettes are encased in a cartridge-like container for protection. As a result, it is often called a disk cartridge.

**disk drive:** The input/output device that writes data on and reads data from a magnetic diskette.

**diskette:** A thin flexible sheet of Mylar, coated with a magnetic-oxide surface on which data is stored in tracks.

**distributed application:** A function which is implemented using co-operating application-processes, characterized by multiple procedures which may be executed in different systems.

**distributed control:** A design approach in which factory or machine control is divided into several subsystems, each managed by a unique programmable controller, yet all interconnected to form a single entity. Various types of communications busses are used to connect the controllers.

**distributed media access control:** A Data Link Function in which the responsibility for media access control is distributed among more than one station.

**distributed processing:** A design approach in which the processing task of a single programmable controller is disbursed among several processors all within the same system.

**documentation:** An orderly collection of recorded hardware and software information concerning the control system. These records provide valuable reference data for installation, debugging, and maintenance of the programmable controller. Examples: Input/Output address assignments, Hardware arrangement (location) diagram, Program printout, etc.

**double precision:** A method of increasing the range of expressible numbers by using multiple bytes or words to represent single numbers.

**down load:** The process of transferring data tables from a network controller to the user memory of an end-device, in order to make that device network-compatible.

**downtime:** The time when a system is not available for production due to required maintenance (scheduled or unscheduled).

**drop line:** A flexible coaxial cable, normally type RG-6U, which usually drops from an overhead tap in the coaxial network. The end of the drop line has the network outlet connector which is used to couple an external device.

**drop line device:** Any external device attached to the coaxial network through a drop line (e.g. RF modem, TV set, audio modulator, etc.).

**DTE:** Data Terminal Equipment. Equipment that includes the data source, data sink, or both.

**dump:** The process of printing out or externally storing the entire contents of the controller's memory. duplex: A communication link in which data transmission can take place in both directions simultaneously.

**dynamic start-up:** That part of the controller start-up procedure in which the system is checked under program control.

-E-

**EAROM:** Electrically Alterable Read Only Memory.

**ECMA:** European Computer Manufacturers' Associated.

**edit:** To deliberately modify a stored progam.

**EEPROM:** Electrically Erasable Programmable Read Only Memory.

**EIA:** Abbreviation for Electronic Industries Association. An agency that sets electrical/electronic standards.

**end-device:** A device located at the end of a particular network drop. An end-device may have direct control over a real application (PC, robot, NC machine), or it may monitor such an application (terminal).

**EPROM:** Erasable Programmable Read Only Memory. A ROM that can be erased with ultraviolet light, and then reprogrammed.

**error control procedure:** That part of a protocol controlling the detection, and possibly the correction, of transmission errors.

**error correcting code (ECC):** A code in which each acceptable expression conforms to specific rules of construction that also define one or more equivalent non-acceptable expressions, so that if certain errors occur in an acceptable expression the results will be one of its equivalents and thus the error can be corrected.

**error detecting code (EDC):** A code in which each expression conforms to specific rules of construction so that if certain errors occur in an expression the resulting expession will not conform to the rules of construction and thus the presence of errors is detected. Synonymous with self-checking code. Exclusive-OR (XOR): A logical operation that has only two inputs, and yields a logic "1" output if either of the two inputs is "1," and a logic "0" output if both inputs are "1" or "0."

**execution:** The performance of a specific operation such as would be accomplished through processing one instruction, a series of instructions, or a complete program.

**execution time:** The time required to perform one specific instruction, a series of instructions, or a complete program. The execution time for a given instruction may vary depending on the outcome of the instruction (i.e. true or false), and the parameters involved.

**Executive Memory:** That portion of the System memory that is responsible for supervisory functions that govern system operation. It allocates and controls system resources such as routines that allow the user to communicate with the processor, and the execution of the application program.

-F-

**file:** A formatted block of data that is treated as a unit (e.g. message file).

**flag:** A programming technique of using a single bit in memory to detect and remember some event. A storage bit.

**flag sequence:** The unique sequence of eight bits (01111110) employed to mark the beginning and ending of a frame.

**floppy disk:** A low cost mass storage medium usually consisting of a thin flexible sheet of Mylar with a magnetic oxide surface. Sometimes called diskette.

**flow chart:** A graphical reprensentation of a definition or method of solution to some task or problem. Example: One might flow chart the step-by-step procedure for load shedding during peak demand hours.

**foreground:** The area in Network Controller memory that contains high-priority application programs.

**forward channels:** In a CATV system, the frequency bands assigned for transmission from the head end into the cable. In a mid-split system, these frequencies are in the range of 150-300 MHz.

**frame:** A sequence of bits which is delimited or otherwise defined by some distinguished bit sequence; most-often applied to sequences processed at the data link layer.

**frame check sequence (FCS):** The field immmediately preceding the ending flag sequence of a frame, containing the bit sequence that provides for the detection of transmission errors by the receiver.

**frequency shift keying (FSK):** A signal modulation technique, in which a carrier frequency is shifted high or low to represent a binary one or zero, respectively. Offers a high degree of noise immunity.

**front-end processor (FEP):** A single-user computer operating system (e.g. DEC RT-11 or 624-30 Translator) that may act as Network Controller. Connected to a host computer, the FEP preprocesses data between the host and the network.

## -G-

**gate:** A circuit having two or more input terminals and one output terminal, where an output is only present when and only when the prescribed inputs are present.

**gateway:** A device (or pair of devices) that connects two or more communications networks. This device might appear to each network as a host on that network. A gateway may transfer messages between network by translating protocols.

**global address:** An 8-bit address field consisting of all ones. A message with this address will be received by all stations on the network.

**global output:** An internal output used for inter-processor communication.

## -H-

**half duplex:** A communication link in which data transmission is limited to one direction at a time.

**handshaking:** Two-way communication between two devices in order to insure successful data transfer. Based on a data ready/data received scheme, one device alerts the other that it is ready to send. The other device signals when ready to receive, and acknowledges when the data is received. Handshaking can be accomplished through hardware using special lines, or through software using special codes.

**handshaking signals:** Special interface lines used for asynchronous (not timed) data transfers.

**hard copy:** A printed document of what is stored in memory. Examples: ladder program listing, Input/Output cross reference, data table contents, etc.

**hardware:** Includes all the physical components of the programmable controller, including peripherals, as contrasted to the programmed software components that control its operation.

**hardwired logic:** Logic control functions that are determined by the way devices are interconnected, as contrasted to prgrammable control in which logic control functions are programmable and easily changed.

**HDLC (High Level Data Link Control):** A communications protocol developed by the International Standards Organization (ISO), which defines procedures for the Data Link and Physical protocol layers (.2 and .1).

**head end:** An electronic control center for a CATV network, Also called a "Hub" for bi-directional (e.g. mid-split) systems

**header:** The control prefix in an asynchronous message frame. The prefix includes source/destination code, messages type, and priority.

**Hexidecimal Number System:** A number system that uses the numerals 0,1,2,3,4,5,6,7,8,9, and the letters A,B,C,D,E,F to represent numbers and codes. Base 16.

**high-level language:** A powerful set of user-oriented instructions, in which each statement may translate into a series of instructions or subroutines in machine language.

**host computer:** A computer attached to a network supplying such services as computation, data base access, or special programs.

301

**-I-**

**information field:** The sequence of bits, occurring between the last bit of the control field and the first bit of the frame check sequence. The information field contents are not interpreted at the link level.

**information rate:** A network measure defined as the amount of data sent per unit time, with units in bits per second.

**input:** Information sent to the processor from connected devices, via some input interface.

**input device:** Any connected equipment that will supply information to the central processing unit such as control devices (e.g. switches, buttons, sensors), or peripheral devices (e.g. CRT, manual programmer). Each type of input device has a unique interface to the processor.

**instruction set:** The set of general-purpose instructions available with a given controller. In general, different machines have different instruction sets.

**intelligent terminal:** An input/output device with built-in intelligence in the form of a microprocessor, and able to perform functions that would otherwise require use of the main memory, and the central processor (e.g. an intelligent CRT programmer).

**interface:** A circuit that permits communication between the central processing unit and a field input or output device. Different devices require different interfaces. Typical interfaces are: ac inputs, ac outputs, dc inputs, dc outputs, analog inputs, analog outputs, etc.

**interframe time fill:** The series of flag sequences transmitted between frames.

**internal output:** A program output that is used strictly for internal purposes (does not drive a field device). It provides interlocking functions like a hardwired control relay, however normally-closed and normally open contacts from an internal output may be used as often as required. Internal outputs also provide a solution to ladder diagram format restrictions imposed by the programming unit. Also called internal storage bit, or internal coil.

**interrupt:** The act of directing a program's execution to a more urgent task.

**I/O:** Abbr. for Input/Output.

**I/O address:** A unique number assigned to each input and output. The address number is used when programming, monitoring, or modifying a specific input or output.

**I/O channel:** A single input or output circuit.

**I/O module:** A plug-in type assembly that contains more than one input or output circut. A module usually contains two or more identical circuits. Normally 2,4,8, or 16 circuits.

**I/O update:** The continuous process of revising each and every bit in the I/O tables, based on the latest results from reading the inputs and processing the outputs according to the control program.

**I/O update time:** The time required to update all local and remote I/O. Also called I/O scan.

**ISO:** Internatinal Standards Organization.

**isolated I/O:** Input and output circuits that are electrically isolated from any and all other circuits of a module. Isolated I/O are designed to allow for connecting field devices that are powered from different sources, to one module.

-J-

**jump:** Change in normal sequence of program execution , by executing an instruction that alters the program counter. Sometimes called a branch. In ladder programs a JUMP (JMP) instruction causes execution to jump forward to a labelled rung. In a high-level language, execution is typically passed to a statement line number or an address label like START.

-L-

**ladder diagram:** An industry standard for representing relay-logic control systems.

**ladder element:** Any one of the elements that can be used in a ladder program. The elements include contacts, coils, shunts, timers, counters, etc.

**ladder program:** A type of control program that uses relay-equivalent contact symbols as instructions (e.g. normally-open, normally-closed contacts, coils).

**ladder matrix:** A rectangular array of programmed contacts, that defines the number of contacts that can be programmed across a row and the number of parallel branches allowed in a single ladder rung.

**language:** A set of symbols and rules for representing and communicating information among people, or between people and machines. The method used to instruct a programmable device to perform various operations. Examples are: boolean, ladder contact symbology, Basic.

**LAN:** Local Area Network.

**latch:** A ladder program output instruction that retains its state even though the conditions which caused it to latch ON may go OFF. A latched output must be unlatched. A latched output will retain its last state (On or Off) if power is removed.

**layered architecture:** A system that breaks network protocol into seven distinct layers. Each layer provides services to the next higher layer, until the highest layer is capable of performing distributed applications.

**leased line:** A private communications line, reserved solely for the customer without interexchange switching.

**LED:** Abbreviation for light-emitting diode. A semiconductor diode, the junction of which emits light when passing a current in the forward direction. LEDs are used as diagnostic indicators on various controller hardware components.

**line printer:** A hard-copy device that prints one line of information at a time.

**liquid crystal display (LCD):** A display device consisting basically of a liquid crystal, hermetically sealed between two glass plates. One type of LCD depends upon a backlighting source. The readout is either dark characters on a dull white background or white characters on a dull black background. LCDs are used on many hand-held programmers.

**local area network:** An ensemble of interconnected processing elements (nodes) which are typically cofined to fall within a radius not exceeding a few miles.

**location:** In reference to memory, a storage position or register identified by a unique address.

**logic:** A process of solving complex problems through the repeated use of simple functions that can be either true or false. The three basic logic functions are AND, OR, and NOT.

**logic diagram:** A drawing which shows AND, OR, and NOT logic symbols, interconnected to graphically describe system operation or control.

**logic level:** The voltage magnitude associated with signal pulses that represent ones and zeros in digital systems.

### -M-

**machine language:** A program written in binary form.

**main memory:** the block of data storage location connected directly to the CPU.
mask: A logical function used to always set certain bits in a word to an established state.

**master station:** A data station that has accepted the nomination to ensure a data transfer to and/or from one or more slave stations. A responder or listener cannot be a master under this definition.

**media access control:** A data link function which determines which station on a data highway shall be enabled to transmit at a given time.

**memory:** That part of the programmable controller where data and instructions are stored either temporarily or semipermanently. The control program is stored in memory.

**memory map:** A diagram showing a system's memory addresses and what programs and data are assigned to each section of memory.

**memory mapped I/O:** A method of interfacing by assigning memory addresses to I/O ports as well as memory. Data is then sent to and read from external devices by simply reading or writing to certain memory locations.

**message:** A group of data and control bits transferred as an entity from a data source to a data sink, whose arrangement of fields is determined by the data source.

**message mode:** A manner of operating a data network by means of message switching.

**memory protect:** The capability of preventing unauthorized memory entries or program changes. Usually provided by a keylock switch, or software access code.

**menu:** In an interactive system (e.g. programming device), a list of system programs, (typically displayed on a viewing screen or other sisplay) of which the user can select which operation he wishes to perform. The menu system provides prompts to guide the user's input and response.

**message switching:** The process of routing messages by receiving, storing, and forwarding complete messages within a data network. Compare with "Packet"

**microprocessor:** A digital electronics-logic package (usually on a single chip), capable of performing the program execution, control, and data processing functions of a central processing unit. The microprocessor usually contains an arithmetic logic unit, temporary storage registers, instruction decode circuitry, a program counter and bus interface circuitry.

**microsecond:** One millionth of a second. $1 \times 10^{-6}$ second or 0.000001 seconds.

**millisecond:** One thousandth of a second. $1 \times 10^{-3}$ second or 0.001 seconds.

**mnemonic codes:** Symbolic names, usually alphanumeric, for referencing instructions, registers, addresses, etc., to eliminate the need for remembering numbers, by substituting meaningful codes.

**modem:** Abbreviation for modulator/demodulator. A device that modulates digital signals and sends them across a telephone line. It also demodulates incoming signals and converts them into digital signals.

**modulation rate:** A network measure defined as the rate at which a signal changes, with units in baud. Differs from information Rate when more than one symbol defines a cell.

**motherboard:** A part of the system hardware that contains the connectors and bus wiring for the systems's circuit boards (e.g. processor boards, memory boards, I/O modules). Also called a backplane.

**MOV:** Metal Oxide Varistor

**multipoint configuration:** A network in which connections are made between more than two statins (end-devices).

**multiprocessing:** Concurrent execution of two or more tasks residing in memory. Available in both RT-11 and RSX-11M operating systems.

**multiplex:** The act of channelling two or more signals to one source, or sharing a system resource.

### -N-

**NAND:** A logical operation that yields a logic "1" output if any input is "0," and a logic "0" if all inputs are "1." The negated AND function. Result of negating the output of an AND gate, by following it with a NOT symbol.

**NBS:** National Bureau of Standards.

**negative logic:** The use of binary logic in such a way that "0" represents the voltage level normally associated with logic 1 (e.g. 0 = +5V, 1 = 0V) Positive logic is more conventional (i.e. 1 = +5V, 0 = 0V).

**network:** A series of points (or devices) connected by some type of communications medium.

**network controller:** A host computer or front-end processor that establishes and maintains the flow of data in the network. It gathers data from the end-devices for presentation to an application program.

**Network Layer:** Layer 3 of the OSI architecture. It provides to the transport-entities independence from routing and switching considerations associated with the establishment and operation of a given network connection. This includes the case where several transmission resources are used in tandem or in parallel. It makes invisible to transport-entities, how the Network Layer uses underlying resources such as data-link-connections, to provide network-connections.

**node:** A point on the network bus, where it connects to a secondary station, at which network messages are received and responses placed.

**noise:** Random, unwanted electrical signals, normally caused by radio waves or electrical or magnetic fields generated by one conductor and picked up by another.

**nonvolatile memory:** A type of memory whose contents are not lost or disturbed if operating power is lost. Examples: EPROM, Core, NOVRAM.

**Nonvolatile Random Access Memory:** A special kind of RAM that will not lose its contents due to a loss of power.

**NOR:** A logical operation that yields a logic "1" output if all inputs are "0" and a logic "0" output if any input is "1." The negated OR function. Result of negating the output of an OR gate, by following it with a NOT symbol. normally-closed contact: 1) A relay contact-pair which is closed when the coil of the relay is not activated, and opens when the coil is activated. 2) A ladder program symbol that will allow logic continuity (flow) if the referenced input is logic "0" when evaluated.

**normally-open contact:** 1) A relay contact-pair which is open when the coil of the relay is not activated, and closes when the coil is activated. 2) A ladder program symbol that will allow logic continuity (flow) if the referenced input is logic "1" when evaluated.

**normal response mode (NRM):** A mode in which a secondary station may only initiate transmission after receiving permission to do so from the primary.

**NOT:** A logical operation that yields a logic "1" at the output if a logic "0" is entered at the input, and a logic "0" at the output if a logic "1" is entered at the input. The NOT, also called the inverter is normally used in conjuction with the AND and OR functions.

**NOVRAM:** See Non-Volatile Random Access Memory.

**NRZI (Non-Return to Zero Inverted):** A self-clocking pulse code used to establish reliable synchronous transmission. The industry standard for serial communications.

**null modem:** A cable that interconnects two RS232C devices by acting as a dummy piece of data communication equipment.

### -O-

**Octal Number System:** A number system that uses eight numeral digits (octal digits), 0,1,2,3,4,5,6,7. Base 8.

**off-line:** Refers to not being in continuous direct communication with the processor; done independent of the processor (e.g. off-line program generation and storage).

**one's complement:** A binary representation in which the MSB of the word is assigned the negative value of its normal weight minus one. Negative numbers can thereby be expressed and numbers can be negated by simply complementing them.

**one-way interaction:** Use of a session wherein one presentation-entity always sends and the other presentation-entity always receives.

**on-line:** Refers to being in continuous communication with the processor while it is running or stopped (e.g. on-line programming).

**open system:** A system which can be interconnected to others according to established standards.

**open system interconnection (OSI) architecture:** A model of system architecture specified in ISO DP 7498 (1980).

**operating system:** A collection of programs that organizes a set of hardware devices into a working unit that people can use.

**optical coupler:** A device that couples signals from one circuit to another by means of electromagnetic radiation, usually infrared or visible . A typical optical coupler uses a light-emitting diode (LED) to convert the electrical signal of the primary circuit into light, and a phototransistor in the secondary to reconvert the light back into an electrical signal. Sometimes referred to as optical isolation.

**optical isolation:** Electrical separation of two circuits with the use of an optical coupler.

**OR:** A logical operation that yields a logic "1" output if one of any number of inputs is "1," and a logic "0 if all inputs are "0."

**OSI Model:** A description of communications functions and services organized as seven layers in such a way as to help promote "open systems interconnection."

**output:** Information sent from the processor to a connected device via some interface. The information could be in the form of control data that will signal some device such as a motor to switch ON or OFF, or vary the speed of a drive. It could also be pure data such as a string of ASCII characters that will generate a report.

**output device:** Any connected equipment that will receive information or instructions from the central processing unit, such as control devices (e.g. motors, solenoids, alarms, etc.) or peripheral devices (e.g. line printers, disk drives, color displays, etc.) Each type of output device has a unique interface to the processor.

**overflow:** The act of exceeding an ALU's or adder's numerical capacity in the positive or negative direction.

**-P-**

**packet:** Data and a sequence of control bits arranged in a specified format and transferred as an entity that is determined by the process of transmission. Compare with "Message."

**packet mode:** A manner of operating a data network by means of packet switching.

**packet switching:** The process of routing and transferring data by means of addressed packets so that a connection is occupied during the transmission of the packet only, and upon completion of the transmission, the connection is made available for transfer of other packets. Packet switching is a combination of multiplexing, segmenting, and routing, whereas message switching involves routing and multiplexing only.

**parallel circuit:** A circuit in which two or more of the connected components or contact symbols in a ladder program, are connected to the same pair of terminals so that current may flow through all the branches, as contrasted with a series connection , where the parts are connected end-to-end so that current flow has only one path.

**parallel communication:** A form of communications in which groups of bits or entire words are transmitted simultaneously.

**parallel transmission:** The simultaneous transmission of a group of bits that constitute a character or a frame of data. Compare "serial transmission."

**parity:** The even or odd characteristic of the number of 1's in a byte or word of memory.

**parity bit:** An additional bit added to each memory word as a means of error detection. See parity check.

**parity check:** A check for the number of 1's in a memory word, by performing an exclusive-OR. If the result is odd, then parity is odd, and if the result is even, then parity is even. The test will be for either odd or even parity, and the wrong result is a parity error. The parity check is to improve reliability on memory and interface transfers.

**partition:** A logical division of main memory. In a multiprogramming system, tasks are executed in a specific partition and in parallel with other tasks.

**peer-to-peer:** a form of communications in which messages are exchanges between entities with comparable functionality in different systems.

**peripherals:** External devices that are connected to the programmable controller (e.g. line printers, cassette recorders, CRT programmers, disk drives etc.)

**permanent virtual circuit (PVC):** A network-facility providing a permanent association between two network-connection-endpoints as specified in CCITT Recommendation X.25. A PVC is analogous to a point-to-point private line; hence, no call setup or call clearing action is required or allowed.

**PID:** Proportional-Intergral Derivative. A mathematical formula that provides a closed loop control of a process. Inputs and outputs are continously variable and typically will be analog signals.

**physical interface:** A shared boundary defined by common physical interconnection characteristics, signal characteristics, and functional characteristics of the interchange circuits.

**physical layer:** Layer 1 of the Open Systems Architecture. It specifies the mechanical and functional characteristics between network nodes. Defined by such standards as RS-232C and RS-422.

**point-to-point configuration:** A network in which a connection is made between two and only two terminal installations.

**polling:** After an instruction message has been sent, the Network Controller sends another message (polls) asking for a result. May be addressed to a single station or to all stations.

**port:** A signal input or output point (e.g. communication port).

**positive logic:** The use of binary logic in such a way that "1" represents a positive logic level (e.g. 1 = +5V, 0 = 0V). This is the conventional use of binary logic.

**power supply:** The unit that supplies the necessary voltage and current to the system circuitry.

**presentation layer:** Layer 6 of the OSI architecture. The purpose of this layer is to represent information to communicating application-entities in a way that preserves meaning while resolving syntax differences. The Presentation layer provides syntax differences. The presentation layer provides the services of data transformation and formatting, syntax selection, and presentation-connections to the application layer.

**primary station:** That part of the data station tht supports the primary control functions of the data link. The primary generates commands for transmission and interprets received responses. Specific responsibilities assigned to the primary include: a) initialization of control signal interchange b) organization of data flow c) actions regarding error control and error recovery functions at the link level. priority: The importance of a device or peripheral in an interrupt system. Each interrupt device has an interrupt service routine. The interrupt will be serviced according to priority in the case of a simultaneous occurrence.

**processor:** See CPU.

**program:** A planned set of instructions stored in memory, and executed in an orderly fashion by the central processing unit.

**programming device:** A device for inserting the control program into memory. The programming device is also used to make changes to the stored program. Most programmers have displays that allow the program to be monitored while the machine or process is in operation. program scan: The time required by the processor to evaluate and execute the control logic. This time does not include the I/O update time. The program scan repeats continuously while the processor is in the run mode.

**programmable controller:** A solid-state control device that can be programmed to control process or machine operation. The programmable controller consists of five basic components(i.e. processor, memory, input/output, power supply, programming device).

**PROM:** Programmable Read Only Memory. A Read Only Memory that can be programmed once, and cannot be altered after that. protocol: A formal definition that describes how communications will take place (e.g. timing considerations, data format, what control signals mean and what they do, the pin numbers for specific functions, the meaning and priority of various messages etc.). Handshaking is a communications protocol.

**PROWAY:** A data highway for process control proposed by IEC TC65/SC65A.

### -R-

**RAM:** Random Access Memory. Commonly referred to as read/write, because it can be written to as well as read from. However, a stricter definition of RAM, is a memory that stores memory in such a way that each bit of information can be stored or retrieved within the same amount of time as any other bit (as contrasted with serial memory, in which data is stored and retrieved in a sequential order).

**Read/Write memory:** : A type of memory which can be read from or written to. Read/Write memory can be altered quickly and easily by merely writing over the part to be changed or inserting a new part to be added. See RAM.

**redundancy:** The number of total characters or bits in a protocol that can be eliminated without a loss in information.

**redundant transmission:** A method of sending each message several times around the network in order to ensure total reception.

**register:** A temporary storage device for various types of information and data. (e.g. timer/counter preset values). In PCs, a register is normally 16 bits wide.

**relay:** An electrically operated device that mechanically switches electrical circuits.

**repeater:** A device whereby signals received over one circuit are automatically repeated in another circuit or circuits, generally in an amplified and/or reshaped form.

**resolution:** The smallest detectable increment of measure. Analog-to-Digital resolution is usually principally limited by the number of bits used to quantize the analog input signal. (e.g. a 12-bit analog-to-digital converter has a resolution of one part in 4096 ($2^{12} = 4096$).

**response:** In data communications, a reply represented in the control field of a response frame. It advises the primary with respect to the action taken by the secondary to one or more commands.

**RFI:** Radio Frequency Interference.

**ROM:** Read Only Memory. A type of memory which permanently stores information (e.g. a math function or a microprogram). A ROM is programmed during fabrication according to the User's requirements, and cannot be reprogrammed.

**routing:** A function within a layer to translate the title or address of an entity into a path by which the entity is to be reached.

**RS232:** An EIA standard, originally introduced by the Bell System, for the transmission of data over a twisted-wire pair, less than 50 ft. in length. It defines pin assignments, signal levels, etc. for receiving and transmitting devices. Other RS-standards cover the transmission of data over distances in excess of 50 ft.

**RS-422:** An EIA Standard for the electrical characteristics of balanced voltage digital interface. Maximum distance-4000ft.

**rung:** A ladder program term, that refers to the programmed instructions that drive one output. A single rung may be only a portion of a complete control program which in turn may be several rungs.

**R/W:** See Read/Write memory.

## -S-

**scan time:** The time required to read all inputs, execute the control program, and update local and remote I/O. This is effectively the time required to activate an output that is controlled by programmed logic.

**SCR:** Abbr. for Silicon Controlled Rectifier. A semiconductor device that functions as an electrically controlled switch for dc loads. A component of dc output circuits.

**scratch pad memory:** A temporary storage area used by the CPU to store a relatively small amount of data used for interim calculations or control. Data that is needed quickly is stored in this area to avoid the access time that would be involved if stored in the main memory.

**secondary station:** That part of the network that executes data link functions as instructed by the primary. A secondary interprets received commands and generates responses for transmission. A slave station in HDLC and ADCCP unbalanced configurations.

**sequencer:** A function of an (N)-entity to provide the (N)-service of delivering data in the same order as its was submitted.

**serial communication:** Type of communication in which bits are transmitted sequentially rather than simultaneously as with parallel communication.

**serial transmission:** The sequential transmission of a group of bits that constitute a character or a frame of data. Compare with "parallel" transmission.

**series circuit:** A circuit in which the components or contact symbols are connected end-to-end, and all must be closed to permit current flow.

**session:** Synonym for session-connection. A cooperative relationship between two application-entities characterizing the commmunication of data between them.

**session-dialogue-service:** A session-service controlling data exchange, delimiting and synchronizing data operation between two presentation-entities.

**session-interaction-unit:** A session-service-data-unit used by presentation-entities to control the transfer of turn in an agreed upon interaction mechanism.

**session layer:** Layer 5 of the OSI architecture. Its purpose is to provide the means for cooperating presentation-entities or organize and synchronize their dialogue and manage their data exchange.

**signal level:** the rms voltage measured during an RF signal peak, usually expressed in microvolts, referred to an impedance of 75 ohms, or in dBmV, the value in decibels with respect to a reference level of 0 dBmV, which is 1 millivolt across 75 ohms.

**significant digit:** A digit that contributes to the precision of a number. The number of significant digits is counted beginning with the digit contributing the most value, called the Most Significant Digit (left-most), and ending with the digit contributing the least value, called the Least Signficant Digit (right-most).

**single-scan:** A supervisory type command initiated by the user while the controller is in the Stop mode. This command causes the control program to be executed for one scan, including I/O update. This troubleshooting function allows step-by-step inspection of occurrences while the machine is stopped.

**slave station:** A station that is selected by a master station to receive data and/or respond with data.

**snubber:** A circuit generally used to suppress inductive loads; it consists of a resistor in series with a capacitor (RC snubber) and/or a MOV placed across the AC load.

**solenoid:** A transducer that converts current into linear motion; it consists of one or more electromagnets that move a metal plunger. The plunger is sometimes returned to its original position after excursion with a spring or permanent magnet.

**solid-state:** Circuitry designed using only integrated circuits, transistor, diodes, etc.; no electro-mechanical devices such as relays are utilized.

**software:** 1) Any written documents associated with the system hardware. 2) Stored instructions (the program).

**splitter:** A passive, 5 - 300 MHz or 800 Mhz band pass device. The device is coupled in-line to a main trunk or branch of a broadband network for splitting the power and the information signal two or more ways on a coaxial network. Splitters always pass through 60Hz power to all outlets.

**start bit:** The first bit sent in an asychronous word transmission. The start bit is for control purposes and does not convey data.

**station:** See Node

**status:** The condition or state of a device (e.g. ON/OFF).

**stop bit:** The last bit in an asynchronous word transmission. The stop bit is for control purposes and does not convey data.

**synchronous transmission:** Transmission in which data is sent at a fixed rate in synchronization with a timing signal or clock pulse. This eliminates the need for start/stop bits.

### -T-

**termination:** 1) The load connected to the output end of a transmission line. 2) A provision for ending a transmission line and connecting to a bus bar or other terminating device (e.g. screw termination).

**time base:** A unit of time generated by the system clock and used by software timer instructions. Normal time bases are 0.01, 0.1, and 1.0 second

**throughput:** The speed at which an application or part of an application is performed. Throughput is dependent on speed, medium, protocol, packet size and amount of data handled by a network.

**time-division multiplex:** A means of supporting more than one date link-connection on single data circuit by enabling data transmission by each station intermittently, generally at regular intervals and by means of an automatic distribution.

**token-passing:** In this message transmission technique a token is passed along the bus and each node has a set amount of time to receive and/or respond to it. Each station must bear a part of the transmission load.

**topology:** The way in which a network is physically structured, such as in a ring, bus, or star configuration.

**transitional contact:** A contact, depending on how it is programmed, will be energized for one program scan every 0 to 1 transition, or every 1 to 0 transition of the referenced coil.

**transmission line:** A physical means of connecting two or more locations to each other for purpose of transmitting and receiving data.

**transmission medium:** The technology used (coaxial cable, fiber optics, etc) in a given transmission line.

**transparency:** A feature, usually provided by an interface unit, which allows the end-device to operate without "knowledge" of network protocol.

**transport layer:** Layer 4 of the OSI architecture. The transport-service relieves its users from any concern with the detailed way in which reliable and cost effective transfer of data is achieved. Transport-functions allow the Network Layer to be composed of more than one communication resource in tandem (e.g. a public packet switched network used in tandem with a circuit-switched network).

**triac:** A semiconductor device that functions as an electrically controlled switch for ac loads. A component of ac output circuits.

**truth table:** A table listing that shows the state of a given output as a function of all possible input combinations.

**TTL:** Abbreviation for Transistor-Transistor Logic. A semiconductor logic family in which the basic logic element is a multiple-emitter transistor. This family of devices is characterized by high speed and medium power dissapation.

**two's complement:** A numbering system used to express positive and negative binary numbers in which the MSB takes on the negative value of its normal weight.


## -U-

**UART:** Abbreviation for Universal Asynchronous Receiver/Transmitter. Interface device for serial/parallel conversion, buffering, and adding check bits.

**unbalanced configuration:** An HDLC or ADCCP configuration involving a single primary station and one or more secondary stations.

**unit interval:** A network measure defined as the shortest code symbol. The reciprocal of baud rate; it should be lond relative to receiver rise and fall times.

**up-load:** the process of transferring end-device data from a network interface to the network controller. Control programs can be up-loaded to be verified against a master copy in memory.

**user memory:** The memory where the application control program is stored

**utility:** A general-purpose program that executes common functions for the central processor.


## -V-

**VA:** Voltamperes

**varistor:** a resistor whose resistance varies proportionately with the voltage applied to it.

**virtual circuit:** An association between two DTE's (a source and a sink) over which data packets are transmitted.

**volatile memory:** A memory whose contents are irretrievable when operating power is lost.


## -W-

**wachdog timer:** A timer that monitors logic circuits controlling the processor. If the watchdog timer which is reset every scan, ever times out, the processor is assumed faulty, and is disconnected from the process.

**word:** The unit number of binary digits (bits) operated on at a time by the central processing unit when it is performing an instruction, or operating on data. A word is usually composed of a fixed number of bits.

**write:** The process of putting information into a storage location.


## -X-

**X.25:** A CCITT recommendation that establishes procedures for the first three layers of network protocol (physical, data link, and network). X.25 provides a link service, but does not provide network control functions.

| AND | OR | A | B | Y |
|---|---|---|---|---|
| A, B → Y | A, B → Y | 1 / 1 / 0 / 0 | 1 / 0 / 1 / 0 | 1 / 0 / 0 / 0 |
| A, B → Y | A, B → Y | 1 / 1 / 0 / 0 | 1 / 0 / 1 / 0 | 0 / 0 / 1 / 0 |
| A, B → Y | A, B → Y | 1 / 1 / 0 / 0 | 1 / 0 / 1 / 0 | 0 / 1 / 0 / 0 |
| A, B → Y | A, B → Y | 1 / 1 / 0 / 0 | 1 / 0 / 1 / 0 | 0 / 0 / 0 / 1 |
| A, B → Y | A, B → Y | 1 / 1 / 0 / 0 | 1 / 0 / 1 / 0 | 1 / 1 / 1 / 0 |
| A, B → Y | A, B → Y | 1 / 1 / 0 / 0 | 1 / 0 / 1 / 0 | 1 / 0 / 1 / 1 |
| A, B → Y | A, B → Y | 1 / 1 / 0 / 0 | 1 / 0 / 1 / 0 | 1 / 1 / 0 / 1 |
| A, B → Y | A, B → Y | 1 / 1 / 0 / 0 | 1 / 0 / 1 / 0 | 0 / 1 / 1 / 1 |

**Logic Diagrams**

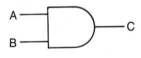

AND Gate

| A | B | C |
|---|---|---|
| 0 | 0 | 0 |
| 0 | 1 | 0 |
| 1 | 0 | 0 |
| 1 | 1 | 1 |

AND Truth Table

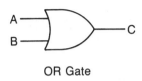

OR Gate

| A | B | C |
|---|---|---|
| 0 | 0 | 0 |
| 0 | 1 | 1 |
| 1 | 0 | 1 |
| 1 | 1 | 1 |

OR Truth Table

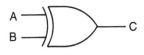

Exclusive-OR
Gate

| A | B | C |
|---|---|---|
| 0 | 0 | 0 |
| 0 | 1 | 1 |
| 1 | 0 | 1 |
| 1 | 1 | 0 |

Exclusive-OR Truth Table

**Ladder Diagrams**

Equivalent
Circuit

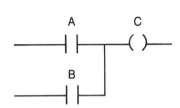

Equivalent
Circuit

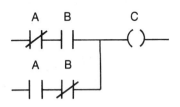

Equivalent
Circuit

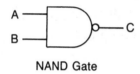

NAND Gate

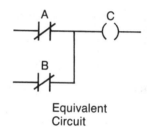

Equivalent
Circuit

| A | B | C |
|---|---|---|
| 0 | 0 | 1 |
| 0 | 1 | 1 |
| 1 | 0 | 1 |
| 1 | 1 | 0 |

NAND Truth Table

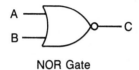

NOR Gate

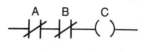

Equivalent
Circuit

| A | B | C |
|---|---|---|
| 0 | 0 | 1 |
| 0 | 1 | 0 |
| 1 | 0 | 0 |
| 1 | 1 | 0 |

NOR Truth Table

| | |
|---|---|
| **ALU:** | Arithmetic Logic Unit |
| **ANSI:** | American National Standards Institute |
| **ASCII:** | American Standard Code for Information Interchange |
| **BCC:** | Block Check Character |
| **BCD:** | Binary Coded Decimal |
| **BCP:** | Byte Or Character Controlled Protocol |
| **CMOS:** | Complimentary Metal Oxide Semiconductor |
| **CPU:** | Central Processing Unit |
| **CRC:** | Cyclic Redundancy Check (For Error Detection) |
| **CRT:** | Cathode Ray Tube |
| **CTS:** | Clear To Send; Handshake Between DTE & DCE |
| **CSMA/CD:** | Carrier Sense Multiple Access with Collision Detection |
| **DLE:** | Data Link Escape (a BCP Control Character) |
| **DMA:** | Direct Memory Access |
| **DSR:** | Data Set Ready: Handshake Between DTE & DCE |
| **DTE:** | Data Terminal Equipment |
| **DTR:** | Data Terminal Ready; Handshake Between DTE & DTE |
| **EAROM:** | Electrically Alterable Read Only Memory |
| **EBCDIC:** | Extended Binary-Coded Decimal Information Code |
| **ECC:** | Error Correction Code |
| **EEPROM:** | Electrically Eraseable Programmable Read Only Memory |
| **EIA:** | Electronic Industries Association |
| **EPROM:** | Eraseable Programmable Read Only Memory |
| **FIFO:** | First-In First-Out |
| **HDLC:** | High (level) Data Link Control |
| **IEEE:** | Institute of Electrical and Electronics Engineers |
| **I/O:** | Input/Output |
| **ISO:** | International Standards Organization |
| **LAN:** | Local Area Network |
| **LCD:** | Liquid Crystal Display |
| **LED:** | Light Emitting Diode |
| **LIFO:** | Last-In First-Out |
| **LRC:** | Longitudinal Redundant Check |
| **LSB:** | Least Significant Bit/Byte |
| **LSI:** | Large Scale Integration |
| **MCR:** | Master Control Relay |
| **MODEM:** | Modulator/Demodulator |
| **MSB:** | Most Significant Bit/Byte |
| **msec:** | millisecond |
| **NOVRAM:** | Non-Volatile Random Access Memory |
| **PC:** | Programmable Controller |
| **PROM:** | Programmable Read Only Memory |
| **RAM:** | Random Access Memory |
| **ROM:** | Read Only Memory |
| **RTS:** | Request To Send |
| **R/W:** | Read/Write |
| **TTL:** | Transistor-Transistor Logic |
| **UART:** | Universal Asynchronous Receiver/Transmitter |
| **$\mu$sec:** | microsecond |
| **VDU:** | Video Display Unit |
| **VPU:** | Video Programming Unit |
| **VRC:** | Vertical Redundancy Check |
| **XFER:** | Transfer |
| **XMIT:** | Transmit |
| **XOR:** | Exclusive OR |

| OCT | PARITY | HEX | ASCII CHAR | CTRL KEYBD EQUIV | ALTERNATE CODE NAMES |
|-----|--------|-----|------------|------------------|----------------------|
| 000 | EVEN | 00 | NUL | @ | NULL, CTRL SHIFT P, TAPE LEADER |
| 001 | ODD | 01 | SOH | A | START OF HEADER,SOM |
| 002 | ODD | 02 | STX | B | START OF TEXT,EOA |
| 003 | EVEN | 03 | ETX | C | END OF TEXT, EOM |
| 004 | ODD | 04 | EOT | D | END OF TRANSMISSION, END |
| 005 | EVEN | 05 | ENQ | E | ENQUIRY,WRU,WHO ARE YOU |
| 006 | EVEN | 06 | ACK | F | ACKNOWLEDGE,RU,ARE YOU |
| 007 | ODD | 07 | BEL | G | BELL |
| 010 | ODD | 08 | BS | H | BACKSPACE,FE0 |
| 011 | EVEN | 09 | HT | I | HORIZONTAL TAB,TAB |
| 012 | EVEN | 0A | LF | J | LINE FEED,NEW LINE,NL |
| 013 | ODD | 0B | VT | K | VERTICAL TAB,VTAB |
| 014 | EVEN | 0C | FF | L | FORM FEED,FORM,PAGE |
| 015 | ODD | 0D | CR | M | CARRIAGE RETURN,EOL |
| 016 | ODD | 0E | SO | N | SHIFT OUT,RED SHIFT |
| 017 | EVEN | 0F | SI | O | SHIFT IN,BLACK SHIFT |
| 020 | ODD | 10 | DLE | P | DATA LINK ESCAPE,DC0 |
| 021 | EVEN | 11 | DC1 | Q | XON,READER ON |
| 022 | EVEN | 12 | DC2 | R | TAPE,PUNCH ON |
| 023 | ODD | 13 | DC3 | S | XOFF,READER OFF |
| 024 | EVEN | 14 | DC4 | T | TAPE,PUNCH OFF |
| 025 | ODD | 15 | NAK | U | NEGATIVE ACKNOWLEDGE,ERR |
| 026 | ODD | 16 | SYN | V | SYNCHRONOUS IDLE,SYNC |
| 027 | EVEN | 17 | ETB | W | END OF TEXT BUFFER,LEM |
| 030 | EVEN | 18 | CAN | X | CANCEL,CNCL |
| 031 | ODD | 19 | EM | Y | END OF MEDIUM |
| 032 | ODD | 1A | SUB | Z | SUBSTITUTE |
| 033 | EVEN | 1B | ESC | [ | ESCAPE,PREFIX |
| 034 | ODD | 1C | FS | \ | FILE SEPARATOR |
| 035 | EVEN | 1D | GS | ] | GROUP SEPARATOR |
| 036 | EVEN | 1E | RS | | RECORD SEPARATOR |
| 037 | ODD | 1F | US | — | UNIT SEPARATOR |
| 040 | ODD | 20 | SP | | SPACE,BLANK |
| 041 | EVEN | 21 | ! | | |
| 042 | EVEN | 22 | " | | |
| 043 | ODD | 23 | # | | |
| 044 | EVEN | 24 | $ | | |
| 045 | ODD | 25 | % | | |
| 046 | ODD | 26 | & | | |
| 047 | EVEN | 27 | ' | | APOSTROPHE |
| 050 | EVEN | 28 | ( | | |
| 051 | ODD | 29 | ) | | |
| 052 | ODD | 2A | * | | ASTERISK |
| 053 | EVEN | 2B | + | | |
| 054 | ODD | 2C | , | | COMMA |
| 055 | EVEN | 2D | - | | MINUS |
| 056 | EVEN | 2E | . | | PERIOD |
| 057 | ODD | 2F | / | | |
| 060 | EVEN | 30 | 0 | | NUMBER ZERO |
| 061 | ODD | 31 | 1 | | NUMBER ONE |
| 062 | ODD | 32 | 2 | | |
| 063 | EVEN | 33 | 3 | | |
| 064 | ODD | 34 | 4 | | |
| 065 | EVEN | 35 | 5 | | |
| 066 | EVEN | 36 | 6 | | |
| 067 | ODD | 37 | 7 | | |
| 070 | ODD | 38 | 8 | | |
| 071 | EVEN | 39 | 9 | | |
| 072 | EVEN | 3A | : | | COLON |
| 073 | ODD | 3B | ; | | SEMI-COLON |
| 074 | EVEN | 3C | < | | LESS THAN |
| 075 | ODD | 3D | = | | |
| 076 | ODD | 3E | > | | GREATER THAN |
| 077 | EVEN | 3F | ? | | |

To transmit control codes, depress"CTRL," then the desired character under keyboard equivalent.

## ASCII Cross Reference (continued)

| OCT | PARITY | HEX | ASCII CHAR | ALTERNATES |
|-----|--------|-----|------------|------------|
| 100 | ODD | 40 | @ | SHIFT P |
| 101 | EVEN | 41 | A | |
| 102 | EVEN | 42 | B | |
| 103 | ODD | 43 | C | |
| 104 | EVEN | 44 | D | |
| 105 | ODD | 45 | E | |
| 106 | ODD | 46 | F | |
| 107 | EVEN | 47 | G | |
| 110 | EVEN | 48 | H | |
| 111 | ODD | 49 | I | LETTER I |
| 112 | ODD | 4A | J | |
| 113 | EVEN | 4B | K | |
| 114 | ODD | 4C | L | |
| 115 | EVEN | 4D | M | |
| 116 | EVEN | 4E | N | |
| 117 | ODD | 4F | O | LETTER O |
| 120 | EVEN | 50 | P | |
| 121 | ODD | 51 | Q | |
| 122 | ODD | 52 | R | |
| 123 | EVEN | 53 | S | |
| 124 | ODD | 54 | T | |
| 125 | EVEN | 55 | U | |
| 126 | EVEN | 56 | V | |
| 127 | ODD | 57 | W | |
| 130 | ODD | 58 | X | |
| 131 | EVEN | 59 | Y | |
| 132 | EVEN | 5A | Z | |
| 133 | ODD | 5B | [ | SHIFT K |
| 134 | EVEN | 5C | \ | SHIFT L |
| 135 | ODD | 5D | ] | SHIFT M |
| 136 | ODD | 5E | ^ | SHIFT N |
| 137 | EVEN | 5F | _ | ,SHIFT O, UNDERSCORE |
| 140 | EVEN | 60 | ACCENT GRAVE | |
| 141 | ODD | 61 | a | |
| 142 | ODD | 62 | b | |
| 143 | EVEN | 63 | c | |
| 144 | ODD | 64 | d | |
| 145 | EVEN | 65 | e | |
| 146 | EVEN | 66 | f | |
| 147 | ODD | 67 | g | |
| 150 | ODD | 68 | h | |
| 151 | EVEN | 69 | i | |
| 152 | EVEN | 6A | j | |
| 153 | ODD | 6B | k | |
| 154 | EVEN | 6C | l | |
| 155 | ODD | 6D | m | |
| 156 | ODD | 6E | n | |
| 157 | EVEN | 6F | o | |
| 160 | ODD | 70 | p | |
| 161 | EVEN | 71 | q | |
| 162 | EVEN | 72 | r | |
| 163 | ODD | 73 | s | |
| 164 | EVEN | 74 | t | |
| 165 | ODD | 75 | u | |
| 166 | ODD | 76 | v | |
| 167 | EVEN | 77 | w | |
| 170 | EVEN | 78 | x | |
| 171 | ODD | 79 | y | |
| 172 | ODD | 7A | z | |
| 173 | EVEN | 7B | [ | |
| 174 | ODD | 7C | ! | VERTICAL SLASH |
| 175 | EVEN | 7D | ] | ALT MODE |
| 176 | EVEN | 7E | ~ | ALT MODE) |
| 177 | ODD | 7F | DEL | DELETE,RUBOUT |

| Table F-1. Powers of Two | |
|---|---|
| $2^n$ | n |
| 1 | 0 |
| 2 | 1 |
| 4 | 2 |
| 8 | 3 |
| 16 | 4 |
| 32 | 5 |
| 64 | 6 |
| 128 | 7 |
| 256 | 8 |
| 512 | 9 |
| 1 024 | 10 |
| 2 048 | 11 |
| 4 096 | 12 |
| 8 192 | 13 |
| 16 384 | 14 |
| 32 768 | 15 |

| Table F-2. Powers of Eight | |
|---|---|
| $8^n$ | n |
| 1 | 0 |
| 8 | 1 |
| 64 | 2 |
| 512 | 3 |
| 4 096 | 4 |
| 32 768 | 5 |
| 262 144 | 6 |
| 2 097 152 | 7 |
| 16 777 216 | 8 |
| 134 217 728 | 9 |
| 1 073 741 824 | 10 |
| 8 589 934 592 | 11 |
| 68 719 476 736 | 12 |
| 549 755 813 888 | 13 |
| 4 398 046 511 104 | 14 |
| 35 184 372 088 832 | 15 |

| Table F-3. Powers of Sixteen | |
|---|---|
| $16^n$ | n |
| 1 | 0 |
| 16 | 1 |
| 256 | 2 |
| 4 096 | 3 |
| 65 536 | 4 |
| 1 048 576 | 5 |
| 16 777 216 | 6 |
| 268 435 456 | 7 |
| 4 294 967 296 | 8 |
| 68 719 476 736 | 9 |
| 1 099 511 627 776 | 10 |
| 17 592 186 044 416 | 11 |
| 281 474 976 710 656 | 12 |
| 4 503 599 627 370 496 | 13 |
| 72 057 594 037 927 936 | 14 |
| 1 152 921 504 606 846 976 | 15 |

Chart of switch symbols for electrical relay diagrams, organized in a bordered grid. Top section titled **SWITCHES** with columns **DISCONNECT**, **CIRCUIT INTERRUPTER**, **CIRCUIT BREAKER**, and **LIMIT** (NORMALLY OPEN, NORMALLY CLOSED, NEUTRAL POSITION, ACTUATED). Subsequent rows include **LIMIT (CONT)** with MAINTAINED POSITION and PROXIMITY SWITCH (CLOSED, OPEN), **LIQUID LEVEL** (NORMALLY OPEN, NORMALLY CLOSED), **VACCUM & PRESSURE** (NORMALLY OPEN, NORMALLY CLOSED), **TEMPERATURE** (NORMALLY OPEN, NORMALLY CLOSED); **FLOW (AIR, WATER)** (NORMALLY OPEN, NORMALLY CLOSED), **FOOT** (NORMALLY OPEN, NORMALLY CLOSED), **TOGGLE**, **CABLE OPERATED (EMERG.) SWITCH**, **PLUGGING**, **NON-PLUG**; **PLUGGING W/LOCK-OUT COIL**, **SELECTOR** (2-POSITION, 3-POSITION), **ROTARY SELECTOR** (NON-BRIDGING CONTACTS, BRIDGING CONTACTS, TOTAL CONTACTS TO SUIT NEEDS); **THERMOCOUPLE SWITCH**, **PUSHBUTTONS** (SINGLE CIRCUIT, DOUBLE CIRCUIT, MUSHROOM HEAD, MAINTAINED CONTACT), **CONNECTIONS, ECT.** (CONDUCTORS: NOT CONNECTED, CONNECTED). Symbol labels visible include DISC, C1, CB, LS, NP, PRS, FS, PS, TAS, FLS, FTS, TGS, COS, PLS, F, R, SS, RSS, LO, TCS, PB.

| CONNECTIONS , ECT. (cont.) | | | CONTACTS | | | | | | |
|---|---|---|---|---|---|---|---|---|---|
| GROUND | CHASSIS OR FRAME NOT NECESSARILY GROUNDED | PLUG AND RECP. | TIME DELAY AFTER COIL | | | | RELAY,ECT. | | THERMAL OVERLOAD |
| | | | NORMALLY OPEN | NORMALLY CLOSED | NORMALLY OPEN | NORMALLY CLOSED | NORMALLY OPEN | NORMALLY CLOSED | |
| GRD | CH | PL / RECP | TR | TR | TR | TR | CR M CON | CR M CON | OL IDL |

| COILS | | | | | | | |
|---|---|---|---|---|---|---|---|
| RELAYS, TIMERS, ECT. | SOLENOIDS,BRAKES,ECT. | | | | THERMAL OVERLOAD ELEMENT | CONTROL CIRCUIT TRANSFORMER | |
| | GENERAL | 2-POSITION HYDRAULIC | 3-POSITION PNEUMATIC | 2-POSITION LUBRICATION | | | |
| CR TR M CON | SOL | SOL 2-H | SOL 3-P | SOL 2-L | OL IOL | HI H3 H2 H4 T X1 X2 | |

| COILS (cont) | | | MOTORS | |
|---|---|---|---|---|
| REACTORS (cont) | | | 3 PHASE MOTOR | DC MOTOR ARMATURE |
| ADJUSTABLE IRON CORE | AIR CORE | MAGNETIC AMPLIFIER WINDING | | |
| X | X | MAX | MTR | MTR A |

| PILOT LIGHTS | | HORN,SIREN, ECT. | BUZZER | BELL | |
|---|---|---|---|---|---|
| LT | PUSH TO TEST | AH | ABU | ABE | |
| R | LT R | | | | |
| LETTER DENOTES COLOR | | | | | |

# Bibliography

Analog-Digital Conversion Handbook.
    Norwood, Massachusetts:
    Analog Devices, 1972.

Application Considerations For Solid-State Controls.
    Allen-Bradley Publication
    SGI-1.1-1976.

Artwick, Bruce A. Microcomputer Interfacing.
    Englewood Cliffs, N.J.:
    Prentice-Hall, 1980.

Bohl, Marilyn. A Guide For Programmers.
    Englewood Cliffs, N.J.:
    Prentice-Hall, 1978.

Boyce, Jefferson C. Digital Logic and Switching Circuits: Operation Analysis.
    Englewood Cliffs, N.J.:
    Prentice-Hall, 1975.

Chilton's Instruments and Control Systems,
    (May 1983), 55-59.

Gould Modicon Division: 184/384
    Users Manual. Gould Inc.
    Modicon Division, 1981.

Hill, Frederick J., Gerald R. Peterson.
    Digital Systems: Hardware Organization and Design.
    New York: John Wiley & Sons, 1973.

Hill, Frederick J., Gerald R. Peterson.
    Introduction to Switching Theory and Logical Design.
    2nd ed.
    New York: John Wiley & Sons, 1974.

Korn, Granino A.
    Minicomputers for Engineers and Scientists.
    New York: McGraw-Hill, 1973.

Landers, George. Xicor Replaces Dip Switches
    and Trimmers with Novram Memories
    Xicor Publication AN-103.
    Milpitas, California 95035

McWhorter, Gene.
    Understanding Digital Electronics.
    Dallas
    Texas Instruments Learning Center, 1978.

Modicon 584; Programmable Controller, Users' Manual.
    Gould Modicon Division, (January 1982), 7-88.

Nagle, Troy H., B.D. Carroll, J. David Irwin.
> An Introduction To Computer Logic.
> Englewood Cliffs, N.J.: Prentice Hall 1975.

National Electrical Code 1981.
> Boston, Massachusetts: National Fire Protection Association, 1981.

Operation/Installation Manual System 6500-450: Durant.
> Eaton Corporation, (November 1981).

PLC-2/30 Programmable Controller: Programming and Operations Manual.
> Allen-Bradley, 1981, 1-1—22-4.

PLC-3 Programmable Controller: Programming and Operations Manual.
> Allen-Bradley, 1981, Publication 1775-801.

Programming Manual: Acramatic Programmable Controller (APC-10).
> Pub. No.7-000- 0420MA.
> Cincinnati, Ohio: Cincinnati Milacron, (January 1982).

Soucek, Branko. Microprocessors and Microcomputers.
> New York: Wiley-Inter Science, 1976.

"SY/MAX: Class 8010 TYPE SPR-200,
> 2100 CRT Programmer," Square D Company
> Instruction Bulletin.

# Index

**Reader Comment Form**

Your comments will assist us in improving our book as well as working on other subjects that may interest you. IPC may use any information you supply in anyway it believes appropriate without incurring any obligation whatever.


Comments:


Return to: International Programmable Controls, Inc.
35 Glenlake Parkway
Suite 445
Atlanta, Georgia 30328